Arturo Buscarino · Luigi F
Mattia Frasca · Gregorio Sciuto

A Concise Guide to Chaotic Electronic Circuits

 Springer

Arturo Buscarino
Luigi Fortuna
Mattia Frasca
Gregorio Sciuto
Dipartimento di Ingegneria Elettrica
    Elettronica e Informatica
University of Catania
Catania
Italy

ISSN 2191-530X   ISSN 2191-5318 (electronic)
ISBN 978-3-319-05899-3   ISBN 978-3-319-05900-6 (eBook)
DOI 10.1007/978-3-319-05900-6
Springer Cham Heidelberg New York Dordrecht London

Library of Congress Control Number: 2014935984

Springer is part of Springer Science+Business Media (www.springer.com)

Preface

Chaos is a remarkable phenomenon occurring in many nonlinear systems, where the deterministic nature of the structure conjugates with the irregularity of the behavior. This means that, despite the fact that the system is described by a set of ordinary differential equations, where all the terms are perfectly known, its behavior is irregular and very sensitive to initial conditions. The first evidence of unpredictability in deterministic systems is found in the work of the mathematician and scientist Henri Poincaré on celestial motion, while the first formulation of chaos in a mathematical model expressed by a set of ordinary differential equations exhibiting chaos is due to the mathematician and meteorologist Edward Lorenz who was studying a model of air motion in the atmosphere and discovered how small variations in the initial values of the variables of his model resulted in divergent weather predictions.

The discovery of Lorenz was then followed by the introduction of other models showing chaotic behavior. However, at the time of these studies an ultimate *experimental* proof of chaos was lacking. This idea inspired the work of Leon O. Chua who designed the first electronic circuit that intentionally behaves in a chaotic way. On the other hand, it was recognized that chaos was already observed in other known circuits developed in the context of electronic oscillators such as the one used by van der Pol (an oscillating electronic valve with a triode), but often classified or eliminated from those circuits as an irregular noise or an unwanted behavior.

The focus of this book is indeed on experiments on chaos and control of chaos. Chaotic circuits are the main subject of the work as well as their characteristic features and chaotic control and synchronization schemes and experiments. In particular, an approach to realize experiments also on systems initially described by a set of dimensionless equations is dealt with: the idea is to design a circuit which obeys to the same equations of the mathematical model so that experiments on it can be performed. The guidelines for the design of such an *equivalent* electronic circuit are discussed and a gallery of chaotic circuits designed and implemented with off-the-shelf components is presented. The book is conceived in such a way that the reader can easily build the selected circuit, verify whether it is properly working and then perform his/her own experiments. On the contrary, the book does not focus on chaos from a theoretical perspective. There are many wonderful books on this subject and the reader is referred to these books for an introduction to chaos.

From the point of view of our study, chaotic circuits are aperiodic electronic oscillators, that is, circuits able to oscillate with irregular waveforms which never repeat themselves. If two variables of a chaotic circuit are reported one against the other on the oscilloscope, converse to what is found for periodic oscillators, that is, an elliptic Lissajous figure, a complex topological structure, which is the signature of the chaotic attractor of the system, appears. The signals generated by chaotic circuits have a number of remarkable features: they oscillate in a long-term unpredictable fashion; their trend is sensitive to small changes in the initial conditions and in parameter values; they have a sharp autocorrelation function; they are uncorrelated with signals coming from different chaotic systems as well as signals coming from different attractors of the same system or different portions of a signal coming from the same attractor.

At the end of this brief introduction, we mention that different applications of chaotic circuits have been, and some are currently, investigated. They refer to the field of nonlinear control of electronic devices and secure communications with the definition of chaos-based encryption techniques or to applications where chaos enhances the device performance, such as the use of chaos to drive sonar sensors in multi-user scenarios or to improve motion control of microrobots.

The book is organized as follows. Chapter 1 introduces four examples of chaotic circuits which have been designed by exploiting specific features of some electronic components. Chapter 2 describes the main fundamental blocks used to design a chaotic circuit obeying a set of mathematical nonlinear ordinary differential equations and the guidelines for the design. Chapter 3 reports a gallery of chaotic circuits, either autonomous or non-autonomous, realized with operational amplifiers and discrete components. Chapter 4 discusses some examples of chaotic circuits implemented by Field Programmable Analog Arrays. Chapter 5 discusses some examples of control and synchronization experiments between two chaotic circuits.

Catania, January 2014

Arturo Buscarino
Luigi Fortuna
Mattia Frasca
Gregorio Sciuto

Contents

Chapter 1
Four Examples of Chaotic Circuits

Abstract In this chapter, four examples of chaotic circuits are given to show the variety of principles underlying chaos generation in electronic circuits. This chapter presents just a sample of the possibilities arising. The analysis of the circuits presented leads to the conclusion that many different mechanisms, not straightforwardly generalizable, can be used for the design of chaotic circuits. The next chapter introduces a procedure, which instead starts from a mathematical model showing chaos and then transfers the mathematical rules of the model into circuit laws of a physical device.

Keywords Chaos · Chaotic circuits · Chua's circuit

1.1 The Chua's Circuit

The Chua's circuit [1, 2] represents the first electronic circuit intentionally designed to behave in a chaotic way. The circuit consists of five elements (two capacitors, an inductor, a resistor, and a nonlinear resistor N_R) connected as in Fig. 1.1. The nonlinear resistor N_R is an active two-terminal element, whose v–i characteristic is represented in terms of a piecewise linear function with three segments (the v–i characteristics are shown in Fig. 1.2). As it can be noticed, the circuit contains three energy-storage elements and one nonlinear element which is also locally active, thus meeting the minimum necessary (although not sufficient) requirements for an autonomous circuit to exhibit chaos. Indeed, the circuit is able to generate a variety of nonlinear phenomena including chaos. For instance, the chaotic attractor known as the double-scroll chaotic attractor is obtained for the parameters $C_1 = 5.5\,\text{nF}$, $C_2 = 49.5\,\text{nF}$, $L = 7.07\,\text{mH}$, $R = 1.428\,\text{k}\Omega$, $G_a = -0.8\,\text{mS}$, $G_b = -0.5\,\text{mS}$, $E = 1\,\text{V}$, and has the typical shape shown in Fig. 1.3.

In the invention of the circuit, Leon O. Chua started from the analysis of two well-known chaotic systems, the Lorenz one and the Rössler one, by noticing that in both

A. Buscarino et al., *A Concise Guide to Chaotic Electronic Circuits*,
SpringerBriefs in Applied Sciences and Technology,
DOI: 10.1007/978-3-319-05900-6_1, © The Author(s) 2014

Fig. 1.1 The Chua's circuit

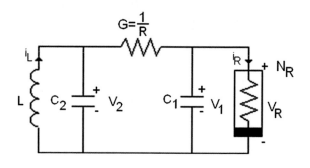

Fig. 1.2 The v–i characteristic of the nonlinear element N_R in the Chua's circuit

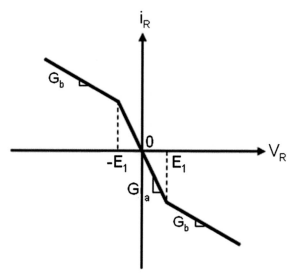

cases chaos was characterized by the presence of at least two unstable equilibrium points and a mechanism generating stable oscillations which are then made unstable by the interaction with the equilibria. Chua aimed to design a circuit with a few elements: three unstable equilibrium points and three energy-storage elements (not of the same type, in order to include a basic mechanism for the birth of oscillations), and followed a systematic approach in drawing all the possible topologies and discarding those not suitable until he obtained the configuration of Fig. 1.1. Once the topology was selected, the choice of the parameters was driven by computer simulations, after which the physical implementation of the circuit successfully showed the onset of chaotic orbits.

It is interesting to note that the same configuration is reached at the end of another path, namely an *evolution* as discussed in [3]. Starting from an LC parallel and adding components increasing the complexity of the circuit, first a configuration with the parallel of an inductor, a capacitor, and a 3-segment PWL resistor is obtained, where the nonlinear resistor allows to obtain stable periodic oscillations. Then, a resistor is

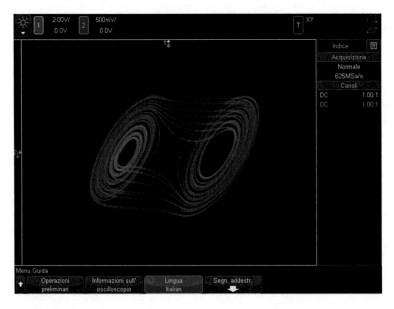

Fig. 1.3 The double-scroll chaotic attractor of the Chua's circuit. Phase plane: V_1–V_2. *Horizontal axis* = 200 mV/div, *vertical axis* = 500 mV/div

added between the nonlinear resistor and the LC subcircuit so that, at the equilibrium, the nonlinear resistor is no more a short circuit and, thus, more than one equilibrium points can be obtained. Finally, another capacitor is added to fulfill the requirement of at least three energy-storage elements in autonomous chaotic circuits.

To conclude this brief introduction to the Chua's circuit, the circuit equations, derived by applying the Kirchhoff's circuit laws, are reported:

$$\begin{aligned}
\frac{dv_1}{dt} &= \frac{1}{C_1}[G(v_2 - v_1) - g(v_1)] \\
\frac{dv_2}{dt} &= \frac{1}{C_2}[G(v_1 - v_2) + i_L] \\
\frac{di_L}{dt} &= -\frac{1}{L}v_2
\end{aligned} \tag{1.1}$$

where v_1, v_2, and i_L represent the voltage across capacitor C_1, the across capacitor C_2, and the current in the inductor L, and $i_R = g(v_R)$ is the nonlinearity of the PWL resistor:

$$g(v_R) = \begin{cases}
G_b v_R + (G_b - G_a)E_1, & \text{if } v_R \leq -E_1 \\
G_a v_R, & \text{if } |v_R| < E_1 \\
G_b v_R + (G_a - G_b)E_1, & \text{if } v_R \geq E_1
\end{cases} \tag{1.2}$$

with G_a being the slope of the inner segment, G_b one of the two outer segments, and $\pm E_1$ the breakpoints. These equations are often rewritten in dimensionless form so that they can be more conveniently studied. This form is reported in Chap. 3, where an equivalent implementation of the Chua's circuit is also discussed.

Fig. 1.4 Scheme of the
single-transistor chaotic
circuit based on the concept of
"disturbance of integration"
[4]

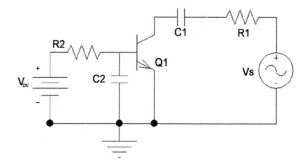

1.2 A Non-autonomous Chaotic Circuit with a Single Transistor

In Sect. 1.1, we discussed how the fundamental element to generate chaos in the Chua's circuit is the 3-segment piecewise linear resistor. Other fundamental chaotic circuits have been designed exploiting the specific features of other electronic components. This section deals with another mechanism used to generate chaos in electronic circuits. Borrowing the terminology from [4], where the concept was introduced, we refer to this mechanism as *disturbance of integration*. The mechanism is illustrated with one example [4], that is, the circuit shown in Fig. 1.4. The circuit makes use of a small number of components: a transistor, two resistors, and two capacitors. In what follows, the components have been chosen as: $R_1 = 1\,k\Omega$, $R_2 = 1\,M\Omega$, $C_1 = 4.7\,nF$, $C_2 = 680\,pF$. Q_1 is a 2N2222A transistor.

In this circuit, the capacitor C_2 is charged through R_2 by a constant source V_{DC}. R_2 is a large resistor ($R_2 = 1\,M\Omega$), which allows to keep constant the value of the DC current flowing into C_2. However, when the transistor C_2 is switched on, it causes a short circuit of C_2. Integration is thus disturbed by the oscillation at the collector of Q_1 that is connected to an RC circuit driven by a sinusoidal oscillator V_s. If the values of the components of the circuit are chosen so that the time constant associated with C_2 is large compared to the period of the sinusoidal oscillator V_s, an antagonistic behavior, leading to chaos, is obtained. For instance, when the circuit is driven by a forcing term equal to $V_s = 7.5\sin(2\pi f t)$ with $f = 3.6\,kHz$, the power supply has been fixed as $V_{DC} = 6\,V$, and the components have been chosen as mentioned above, the chaotic attractor, shown in Fig. 1.5, is obtained. We mention that the same mechanism is also implemented through the circuit shown in Fig. 1.6 (introduced in [4]).

1.3 The Chaotic Colpitts Circuit

Another example of chaotic circuits obtained by using only one transistor is provided by the chaotic Colpitts oscillator [5]. The circuit is obtained starting from the family of Colpitts oscillators. The term Colpitts oscillator, in fact, is used to indicate several

Fig. 1.5 Experimental results of the V_s–V_b for the non-autonomous chaotic circuit with a single transistor

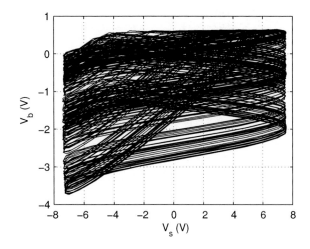

Fig. 1.6 A second example of single-transistor chaotic circuit based on the concept of disturbance of integration [4]. Components are: $R_1 = 1\,\text{k}\Omega$, $R_2 = 1\,\text{M}\Omega$, $C_1 = 4.7\,\text{nF}$, $C_2 = 1.1\,\text{nF}$. Q_1 is a 2N2222A transistor

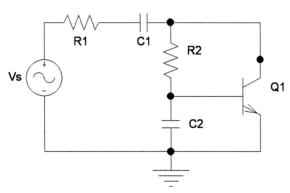

possible configurations in which the combination of an inductance, two capacitors, some resistances, and a nonlinear element (in general, a transistor) is used to generate periodic waveforms. Peculiarities of this type of oscillators are simplicity and robustness. Recently, it has been demonstrated that for some parameter values, a chaotic behavior can also be obtained, thus showing that it is possible to design Colpitts oscillators able to generate high-frequency chaotic signals. The circuital equations of the chaotic Colpitts circuit, whose schematic is reported in Fig. 1.7, are represented by the following state equations:

$$\begin{cases} \frac{dV_1}{dt} = \frac{1}{C_1}[i_L + \beta i_B] \\ \frac{dV_2}{dt} = \frac{1}{C_2}[\frac{V_{cc}-V_2}{R_2} + i_L + i_B] \\ \frac{di_L}{dt} = \frac{1}{L}[V_{cc} - V_1 - V_2 - R_1 i_L] \end{cases} \tag{1.3}$$

where $i_B = 0$ if $V_{BE} \leq V_{TH}$, $i_B = \frac{V_{BE}-V_{TH}}{R_{ON}}$ if $V_{BE} > V_{TH}$, V_{TH} being the transistor threshold voltage $V_{TH} \approx 0.75\,\text{V}$, R_{ON} is the small-signal on-resistance of the

Fig. 1.7 Electrical scheme of
the chaotic Colpitts circuit.
Components $R_1 = 35\,\Omega$,
$R_2 = 500\,\Omega$, $C_1 = 54\,nF$,
$C_2 = 54\,nF$, $L = 98.5\,\mu H$,
2N2222 BJT transistor, $V_{cc} = 5\,V$, $V_{ee} = -5\,V$

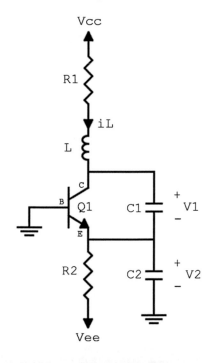

Fig. 1.8 A picture of the
implemented chaotic Colpitts
circuit

base-emitter junction of the transistor and β is the forward current gain. In the experimental setup, these parameters are estimated as $\beta = 200$ and $R_{ON} = 100\,\Omega$. We also notice that the Colpitts oscillator requires a small set of components with a cost that in the limit case can be almost zero. In fact, in our implementation of Fig. 1.8, the inductor L has been extracted from a power supply stage of an out-of-order personal computer and the other components from other recycled electronic boards. The chaotic attractor generated by the Colpitts oscillator is shown in Fig. 1.9.

Fig. 1.9 Experimental
results: projection on the plane
V_2–V_1 of the Colpitts attractor.
Horizontal axis 200 mV/div;
vertical axis 500 mV/div

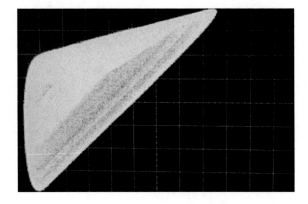

Fig. 1.10 Electrical scheme
of the chaotic circuit designed
by Saito [6]

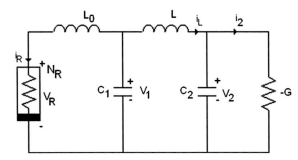

1.4 Chaotic Circuits Based on Hysteretic Components

As a further mechanism to generate chaos, we discuss in this section circuits that include hysteretic components. We first briefly discuss the topology introduced by Saito [6] and then show how the same principle is used to generate chaos in a circuit including a ferroelectric device exhibiting hysteresis.

The electrical scheme of the circuit designed by Saito is shown in Fig. 1.10. It consists of two inductors, two capacitors, one negative resistance, and one nonlinear resistor. As in the Chua's circuit, the nonlinear resistor N_R has a 3-segment piecewise linear i_R–v_R characteristics (the nonlinearity is shown in Fig. 1.11). The peculiarity of the circuit is that for small L_0 the nonlinear resistor operates as an element with hysteresis. The inner segment of slope $-r_1$ is not interested in the chaotic trajectory which only passes through the outer segments having slope equal to r_1. In the limit of small L_0 the trajectory jumps from one outer segment to the other one as schematically depicted in Fig. 1.11. This generates a switching mechanism between two regions where the dynamics are regulated by linear equations. When the switching becomes irregular, a chaotic behavior is obtained. In fact, the circuit exhibits bifurcations from periodic orbits to tori and then to chaotic attractors and is also able to generate hyperchaos, that is, a regime in which two Lyapunov exponents are positive.

Fig. 1.11 Nonlinearity used
in the Saito circuit

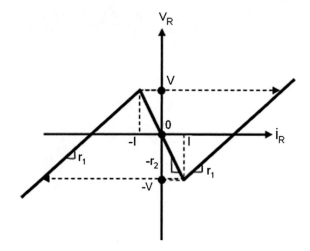

The Saito circuit is governed by the following equations:

$$\begin{aligned}
\frac{dv_1}{dt} &= -\frac{1}{C_1}(i_L + i_R)\\
\frac{dv_2}{dt} &= \frac{1}{C_2}(Gv_2 + i_2)\\
\frac{di_L}{dt} &= \frac{1}{L}(v_1 - v_2)\\
\frac{di_R}{dt} &= \frac{1}{L_0}(v_1 - f(i_R))
\end{aligned} \tag{1.4}$$

where the nonlinearity is given by

$$f(i_R) = \begin{cases}
r_1(i_R - I) - V, & \text{if } i > I\\
-r_2 i_R, & \text{if } |i| < I\\
r_1(i_R + I) + V, & \text{if } i < -I
\end{cases} \tag{1.5}$$

with $I = V/r_2$ and V being a constant.

A very important conclusion to the study of the Saito circuit is that the principle leading to chaos is quite general. In fact, the idea of using hysteresis for generating chaos can be applied to design chaotic circuits with components that are intrinsically characterized by hysteresis. This is the case of the circuit including a nonlinear ferroelectric component discussed in [7]. In this work, the ferroelectrics constitute the medium interposed between the two plates of a capacitor, and is obtained by successive vapor deposition of Strontium, Tantalum, and Bismuth on platinum substrates in small areas. The device exhibits a nonlinear hysteretic behavior as characterized by estimating the output voltage by using a Sawyer-Tower configuration. This ferroelectric component is the core of the circuit which is shown in Fig. 1.12 and operates according to the principle of the Saito circuit. The chaotic attractor is shown in Fig. 1.13.

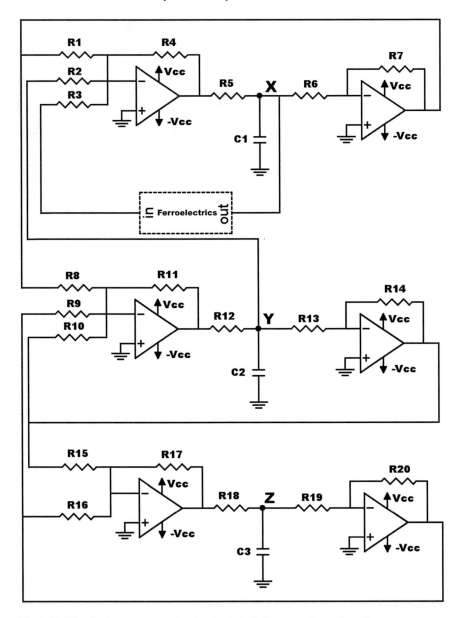

Fig. 1.12 Electrical scheme of a chaotic circuit including a nonlinear ferroelectric component. Components: $R_1 = 100\,\mathrm{k\Omega}$, $R_2 = 100\,\Omega$, $R_3 = 60\,\mathrm{k\Omega}$, $R_4 = 100\,\mathrm{k\Omega}$, $R_5 = 1\,\mathrm{k\Omega}$, $R_6 = 100\,\mathrm{k\Omega}$, $R_7 = 100\,\mathrm{k\Omega}$, $R_8 = 100\,\mathrm{k\Omega}$, $R_9 = 100\,\mathrm{k\Omega}$, $R_{10} = 100\,\mathrm{k\Omega}$, $R_{11} = 100\,\mathrm{k\Omega}$, $R_{12} = 1\,\mathrm{k\Omega}$, $R_{13} = 100\,\mathrm{k\Omega}$, $R_{14} = 100\,\mathrm{k\Omega}$, $R_{15} = 1\,\mathrm{k\Omega}$, $R_{16} = 375\,\Omega$, $R_{17} = 1\,\mathrm{k\Omega}$, $R_{18} = 1\,\mathrm{k\Omega}$, $R_{19} = 100\,\mathrm{k\Omega}$, $R_{20} = 100\,\mathrm{k\Omega}$, $C_1 = 33\,\mathrm{nF}$, $C_2 = 4.71\,\mathrm{nF}$, $C_3 = 33\,\mathrm{nF}$, $V_{cc} = 9\,\mathrm{V}$

Fig. 1.13 Projection in the
phase plane x–y of the chaotic
attractor of the circuit of
Fig. 1.12

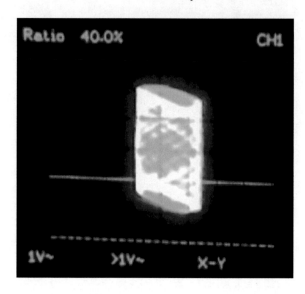

References

1. Madan RN (1993) Chua's circuit: a paradigm for chaos. World Scientific series on nonlinear sciences, series B. World Scientific Publishing, Singapore
2. Fortuna L, Frasca M, Xibilia MG (2009) Chua's circuit implementations: yesterday, today, tomorrow. World Scientific, Singapore
3. Kennedy MP (1993) Three steps to chaos—part I: evolution. IEEE Trans Circuits Syst I 40(10):640–656
4. Lindberg E, Murali K, Tamasevicius A (2005) The smallest transistor-based nonautonomous chaotic circuit. IEEE Trans Circuits Syst II 52(10):661–664
5. Kennedy MP (1994) Chaos in the Colpitts oscillator. IEEE Trans Circuits Syst I Fundam Theor Appl 41:771–774
6. Saito T (1991) Reality of chaos in four-dimensional hysteretic circuits. IEEE Trans Circuits Syst I 38(12):1517–1524
7. Fortuna L, Frasca M, Graziani S, Reddiconto S (2006) A chaotic circuit with ferroelectric nonlinearity. Nonlinear Dyn 44(1–4):55–61

Chapter 2
From the Mathematical Model to the Circuit

Abstract Starting from the mathematical model of a nonlinear system, it is always possible to realize an electronic circuit, which is equivalent to the mathematical model, in the sense that it obeys to the same set of equations. In this chapter, the approach for designing the electronic circuit, equivalent to a given mathematical model, is illustrated.

Keywords Chaotic circuits · Design guidelines · Operational amplifier

2.1 Building Blocks

This section summarizes the basic blocks needed for the realization of an electronic circuit equivalent to a nonlinear system.

2.1.1 Operational Amplifier

The main building block for nonlinear circuits is the operational amplifier (OP-AMP). OP-AMPs are electronic devices important for a wide range of applications [1]. They are characterized by two differential inputs V_+ and V_- and one output V_{out}. The circuital symbol used is shown in Fig. 2.1, where the power supplies are indicated as $-V_{cc}$ and V_{cc}. The transfer characteristic of the OP-AMP from input to the output is nonlinear and can be expressed as follows:

$$V_{out} = f(v_d) = \begin{cases} -E_{sat} & v_d \leq -\frac{E_{sat}}{A_v} \\ A_v v_d & \frac{-E_{sat}}{A_v} < v_d < \frac{E_{sat}}{A_v} \\ E_{sat} & v_d \geq \frac{E_{sat}}{A_v} \end{cases} \qquad (2.1)$$

A. Buscarino et al., *A Concise Guide to Chaotic Electronic Circuits*,
SpringerBriefs in Applied Sciences and Technology,
DOI: 10.1007/978-3-319-05900-6_2, © The Author(s) 2014

Fig. 2.1 Symbol of the
operational amplifier

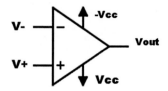

where:

- E_{sat} is the voltage value at which the output of the operational amplifier saturates. It depends on the internal circuitry design of the device and on the voltage supply used. The region in which $V_{out} = A_v v_d$ is defined as the linear region;
- $v_d = V_- - V_+$, is the voltage between the two terminals V_+ and V_-.

In the ideal case, the device has high input impedance, low output impedance and a high voltage gain A_v. As a consequence of the high input impedance, no current flows into or out of the input terminals. An operational amplifier can be integrated in a single chip and used to implement several types of mathematical operations according to the specific configuration. We will discuss the following configurations needed in the design procedure:

1. inverting configuration;
2. non-inverting configuration;
3. algebraic adder;
4. RC integrator;
5. Miller integrator.

2.1.2 Inverting Configuration

In Fig. 2.2 the inverting configuration is shown. An inverting amplifier uses negative feedback to amplify the input voltage while changing its sign. The output v_{out} is related to the input v_{in} through the following equation:

$$V_{out} = -\frac{R_2}{R_1} V_{in} \qquad (2.2)$$

where the gain is fixed by the ratio between R_2 and R_1.

This relationship can be derived by taking into account that the current i flowing into the resistor R_1 is given by:

$$i = \frac{V_{in} + v_d}{R_1} \qquad (2.3)$$

Fig. 2.2 Inverting configuration of the operational amplifier

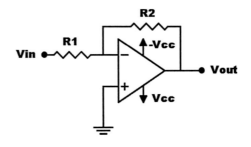

Fig. 2.3 Inverting adder configuration of the operational amplifier

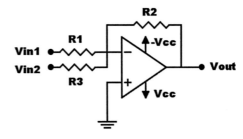

Since no current flows in the negative input terminal, because of the high impedance of this node, the current in R_2 is the same in R_1, and thus:

$$V_{\text{out}} = -R_2 i + v_d \tag{2.4}$$

It is possible to derive that:

$$V_{\text{out}} = -\frac{R_2}{R_1} V_{\text{in}} - \left(\frac{R_2}{R_1} + 1\right) v_d \tag{2.5}$$

and considering that, usually, the device works in the linear region, one gets:

$$\frac{V_{\text{out}}}{V_{\text{in}}} = -\frac{\frac{R_2}{R_1} A_v}{A_v + \left(\frac{R_2}{R_1} + 1\right)} \tag{2.6}$$

In the limit of large gain, $A_v \to \infty$, the relationship (2.2) is obtained. An inverting adder is built from this basic configuration by considering more than one input. The scheme is shown in Fig. 2.3. In the limit of large gain the output is given by:

$$V_{\text{out}} = -\frac{R_2}{R_1} V_{\text{in1}} - \frac{R_2}{R_3} V_{\text{in2}} \tag{2.7}$$

Fig. 2.4 Non-inverting
configuration of the
operational amplifier

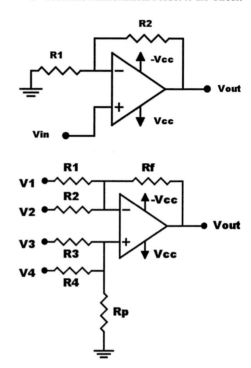

Fig. 2.5 Algebraic adder
configuration of the
operational amplifier

2.1.3 Non-inverting Configuration

In Fig. 2.4 the non-inverting configuration is shown. A non-inverting amplifier realizes an amplification of the input voltage. By following considerations similar to
what taken into account for the inverting configuration, it can be demonstrated that,
in this case, the output V_{out} is given by:

$$V_{out} = \left(\frac{R_2}{R_1} + 1 \right) V_{in} \tag{2.8}$$

2.1.4 Algebraic Adder Configuration

The equations of a nonlinear system often include the sum of more than one term.
So, the need of implementing a mathematical operation like an algebraic sum arises.
To do this, an algebraic adder circuit configuration is used. The scheme is reported
in Fig. 2.5. It implements the following mathematical operation:

Fig. 2.6 RC integrator

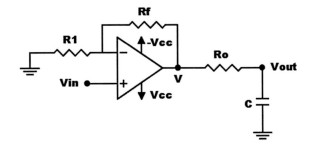

$$V_{\text{out}} = -\frac{R_f}{R_1}V_1 - \frac{R_f}{R_2}V_2 - \frac{R_f}{R_3}V_3 - \frac{R_f}{R_4}V_4 \tag{2.9}$$

The value of the resistor R_p is fixed in order to satisfy the following equation (sometime referred as the *gain rule*):

$$\frac{1}{R_1} + \frac{1}{R_2} = \frac{1}{R_3} + \frac{1}{R_4} + \frac{1}{R_p} \tag{2.10}$$

which, essentially, requires that the sum of the conductances at the inverting terminal of the operational amplifier equals that at the negative terminal. Under this assumption, the output of the circuit is given by the following equation:

$$V_{\text{out}} = \sum_i A_i V_i \tag{2.11}$$

with $A_i = \frac{R_f}{R_i}$. The output depends on each single input by means of only the associated input resistor and not of the other resistors, which is very convenient from the designer perspective. We notice that, when satisfying the gain rule results in a negative value of R_p, another resistance connected to ground should be added at the negative input of the OP-AMP.

2.1.5 RC Integrator

Another important mathematical operation required in the derivation of equivalent electronic circuits is realized by the integrator configuration that exploits the properties of the operational amplifier in the linear region. The configuration is shown in Fig. 2.6.

Assuming that the node V_{out} is connected to an high impedance, the current flowing in the resistor R_o can be considered equal to that in the capacitor. The current flowing into the resistor R_o is:

Fig. 2.7 Miller integrator

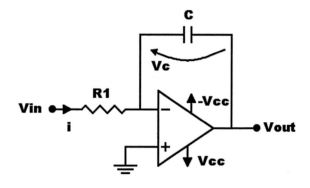

$$i = \frac{V - V_{out}}{R_o} \qquad (2.12)$$

On the other hand, taking into account the relationship between current and voltage across the capacitor, one gets:

$$i = C\frac{dV_{out}}{dt} \qquad (2.13)$$

$$CR_o\dot{V}_{out} = -V_{out} + AV_{in} \qquad (2.14)$$

Thanks to the relationship in Eq. (2.14) and provided that the input V_{in} is appropriately selected, a base block implementing a first-order generic differential equation of the type:

$$\dot{x} = k(-x + f(x, t)) \qquad (2.15)$$

may be realized with $k = \frac{1}{CR_o}$.

2.1.6 Miller Integrator

The Miller integrator is another circuit which allows to obtain an output corresponding to the integral of the input signal. The scheme of the circuit is shown in Fig. 2.7.

This configuration is similar to the inverting configuration of Fig. 2.2, where the resistance R_2 has been replaced by the capacitor C. Considering an ideal operational amplifier, the current in the resistor R_1 and that in the capacitor C are equal, the voltage difference between the inverting and non-inverting terminals is equal to zero and the inverting terminal is connected to virtual ground. The current flowing in R_1 is:

Fig. 2.8 Miller integrator
with feedback resistor

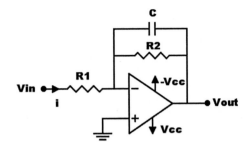

$$i = \frac{V_{in}}{R_1} \qquad (2.16)$$

and, since

$$i = C\frac{dV_c}{dt} \qquad (2.17)$$

and $V_{out} = -V_c$, one has:

$$CR_1\dot{V}_{out} = -V_{in}. \qquad (2.18)$$

The drawback of the circuit of Fig. 2.7 is that it can easily go into saturation, due to low frequency noise or offsets; in fact, if the frequency of the noise tends to zero, the reactance of the capacitor C tends to infinity and so the capacitor becomes an open circuit and its amplification is therefore infinite, so reaching the saturation.

To avoid this, a resistor is inserted in parallel to the capacitor, as shown in Fig. 2.8, so that the maximum gain of the operational amplifier is limited to the value $A_v = -\frac{R_2}{R_1}$. The resistance R_2 must be selected so that at the working frequency of the integrator the presence of the resistance is negligible (that is, the resistance is larger than) with respect to the reactance of the capacitor:

$$\frac{1}{\omega C} \gg R_2 \qquad (2.19)$$

where $\omega = 2\pi f$ and f is the working frequency of the integrator.

2.1.7 The Analog Multiplier AD633

Many chaotic circuits have polynomial nonlinearities or products of state variables. Electronic realization of the product operation may be carried out with the AD633 analog component. This is a low-cost multiplier with the functional block diagram of Fig. 2.9.

Fig. 2.9 Functional block
diagram of the analog
multiplier AD633

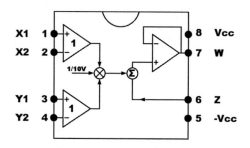

The output of AD633 is related to its input through:

$$W = \frac{(X1 - X2)(Y1 - Y2)}{10\,\text{V}} + Z. \tag{2.20}$$

2.1.8 PWL Approximation of Nonlinearities

The approach based on analog multipliers may be expensive for higher order nonlinearities, thus requiring the cascade of two or more multipliers, or when many chaotic circuits have to be realized. Alternatively, it is possible to use an approach based on piecewise linear (PWL) functions. The idea is to derive a PWL approximation of the nonlinearity to be implemented, to realize a circuitry with the PWL given characteristics, and then to use it instead of the multipliers in the implementation of the circuit. In fact, it has been demonstrated that a wide class may be approximated by using only piecewise linear functions [2]. The design of such PWL functions, and in particular the number of segments, depends on the desired accuracy and on the dynamical range in which the approximation holds. The circuitry implementing the PWL functions is realized with a few components, including diodes to implement the segment breakpoints.

Two examples of circuits that will then be used in Chap. 3 are given here. They both properly work with inputs in the dynamic range of ±3 V. A circuit whose output is the square of the input signal is reported in Fig. 2.10, while Fig. 2.11 implements a PWL approximation of a circuit whose output is the cube of the input signal. For low-cost implementations, these two circuits are more convenient than circuits based on the multiplier $AD633$ since they contain only diodes, resistors, and operational amplifiers.

2.1.9 Negative Resistance

In Fig. 2.12 the symbol and an electronic circuit that can be used as a negative resistance are shown. A brief explanation of the principles of the circuit is reported below. Considering the operational amplifier with ideal characteristics:

$$V_{\text{nr}} = R_1 i + V_o \tag{2.21}$$

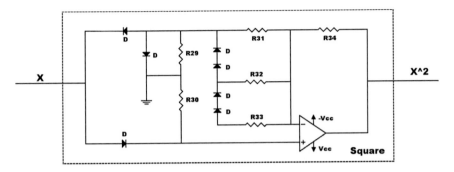

Fig. 2.10 Circuital implementation of the square function. Components: $R_{29} = 10\,\text{k}\Omega$, $R_{30} = 10\,\text{k}\Omega$, $R_{31} = 10\,\text{k}\Omega$, $R_{32} = 10\,\text{k}\Omega$, $R_{33} = 4\,\text{k}\Omega$, $R_{34} = 30\,\text{k}\Omega$, $1N4148$ Diode, $V_{\text{cc}} = 9\,\text{V}$

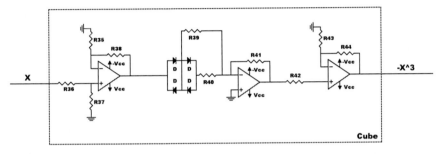

Fig. 2.11 Circuital implementation of the cube function. Components: $R_{35} = 200\,\text{k}\Omega$, $R_{36} = 200\,\text{k}\Omega$, $R_{37} = 100\,\text{k}\Omega$, $R_{38} = 100\,\text{k}\Omega$, $R_{39} = 12\,\text{k}\Omega$, $R_{40} = 2\,\text{k}\Omega$, $R_{41} = 15\,\text{k}\Omega$, $R_{42} = 10\,\text{k}\Omega$, $R_{43} = 10\,\text{k}\Omega$, $R_{44} = 70\,\text{k}\Omega$, $1N4148$ Diode, $V_{\text{cc}} = 9\,\text{V}$

Fig. 2.12 Symbol and circuital implementation of a negative resistance

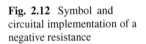
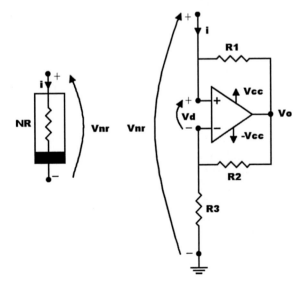

Fig. 2.13 The time-delay block. Schematics of the Sallen-Key low-pass active filter implementing a low-pass Bessel filter. Component values: $R_1 = 10\,\text{k}\Omega$, $R_2 = 10\,\text{k}\Omega$, $C_1 = 10\,\text{nF}$, $C_2 = 22\,\text{nF}$, $V_{cc} = 9\,\text{V}$

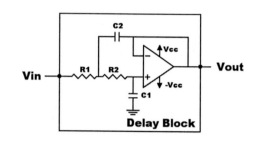

and

$$V_{nr} = V_o \frac{R_3}{R_2 + R_3} + v_d. \tag{2.22}$$

Considering $v_d = 0$, from the relationship (2.22) we get

$$V_o = \frac{R_2 + R_3}{R_3} V_{nr}. \tag{2.23}$$

Substituting V_o in Eq. (2.21) we obtain:

$$V_{nr} = R_1 i + \frac{R_2 + R_3}{R_3} V_{nr} \tag{2.24}$$

from which, considering $R_1 = R_2$, we get:

$$V_{nr} = -R_3 i \tag{2.25}$$

which represents the $i - v$ relationship of the negative resistance.

2.1.10 Time-Delay Block

Another element which is worth to discuss is the time-delay block: in fact, there are nonlinear systems where the presence of a time-delay is fundamental to have chaos [3, 4]. The time-delay block may be implemented by using a cascade of elementary time-delay blocks, where each elementary block is implemented by using a low-pass second-order Bessel filter. Each filter is implemented through the Sallen-Key topology shown in Fig. 2.13 and it is characterized by the following transfer function:

$$H(s) = \frac{1}{1 + C_1(R_1 + R_2)s + C_1 C_2 R_1 R_2 s^2}. \tag{2.26}$$

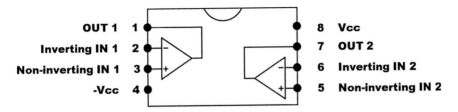

Fig. 2.14 TL082 functional block diagram

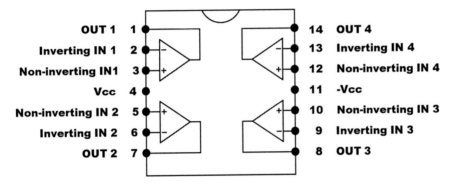

Fig. 2.15 TL084 functional block diagram

The values of the filter components have been chosen in order to realize a Bessel filter with 3 dB frequency equal to $f_c \approx 1$ kHz and taking into account off-the-shelf component values. The time-delay introduced by this filter in the band up to 3 dB can be calculated as $\tau = C_1(R_1 + R_2) = 0.219$ ms. Larger delays are realized by taking into account a cascade of n filters.

2.1.11 General Purpose Amplifiers

TL082 and TL084 are low cost, high speed operational amplifiers. They require low supply voltage, yet maintaining a large gain bandwidth. For this reason, the circuits discussed in the next chapter use such devices. The difference between TL082 and TL084 relies in the number of operational amplifiers integrated in the chip, as apparent from the functional block diagrams of Figs. 2.14 and 2.15. The TL082 has only two operational amplifiers while the TL084 has four.

2.2 Methodology

In this section the methodology to design equivalent electronic circuits starting from a mathematical model is discussed. The mathematical model is expressed in the form

$$\dot{x} = f(x, t) \tag{2.27}$$

where $x \in R^n$, $f : R^n \rightarrow R^n$ and $\dot{x} = \frac{dx}{dt}$. The system in Eq. (2.27) is rewritten as:

$$\dot{x} = -x + Ax + g(x) \tag{2.28}$$

where the linear part and the nonlinear part of the system are emphasized. To each equation of the set of Eq. (2.28) a RC integrator equation in the following form may be associated:

$$\dot{x}_i = -x_i + h_i(x) \tag{2.29}$$

where $h_i(x) = \sum_j a_{ij} x_j + g_i(x)$. To implement $h_i(x)$ an algebraic adder that realizes $\sum_j a_{ij} x_j$, and $g_i(x)$ and then a nonlinear block that implement $g_i(x)$ are needed.

The design of an equivalent nonlinear circuit starting from a set of ordinary differential equations follows three steps.

First step
Any mathematical model, which is the basis of a dynamical system, has a number of state variables following a particular temporal trend. These trends are within a range which depends on the model and its parameters.

In order to implement the model with an electronic circuit using standard circuital components (resistors, capacitors, operational amplifiers, etc.), it is important that the oscillations of the state variables are confined within the limits imposed by the voltage supplies powering the operational amplifiers (otherwise, undesired saturations may be reached). To establish the specific power supply voltage, it is necessary to examine the model using simulation tools. If there are state variables or linear or nonlinear combinations of these which are outside the range of the limits imposed by the voltage supplies, the system in Eq. (2.29) must be rescaled in amplitude, using a transformation of the type:

$$\mathbf{X} = \mathbf{kx} \tag{2.30}$$

where $\mathbf{X} \in R^n$, $\mathbf{x} \in R^n$, and $\mathbf{k} = \begin{bmatrix} k_1 & 0 & \cdots & 0 \\ 0 & k_2 & \cdots & 0 \\ \vdots & \vdots & \ddots & \vdots \\ 0 & 0 & \cdots & k_n \end{bmatrix}$ where $k_1, k_2, \ldots, k_n$ are real

quantities. To check the limits of the oscillations and those of the rescaled variables numerical simulations are performed. The numerical integration of the mathematical model is done by using a solver of ordinary differential equations, as for instance one of those provided in the software *MATLAB®*. In this first phase, the oscillation range and the operating frequencies are characterized. However, if the state variables oscillate outside physically realizable voltage limits, the system has to be suitably scaled in amplitude through the mathematical transformation (2.30) by appropriately selecting $\mathbf{k}$.

The time variable of the dynamical system can also be rescaled by defining a new time variable as $\tau = \kappa t$. According to this Eq. (2.27) is rewritten as

$$\frac{dx}{d\tau} = \kappa f(x, \tau) \tag{2.31}$$

By comparing Eq. (2.31) with Eq. (2.15), we derive that the time scaling factor in the RC integrator is fixed by the product of capacitor C and resistor R_o as $\kappa = \frac{1}{CR_o}$. When n RC blocks are used in the circuit to implement the n state variables, all the capacitors and resistors of the blocks are chosen to match the same scaling factor. The same considerations apply to the Miller integrator. The temporal rescaling is introduced to reduce the observation time of the electrical waveforms.

Second step
In this step, the circuit is designed and simulated according to the observations made during the previous analysis, checking the feasibility of the designed circuit. $PSPICE^{®}$, $LT Spice$, and other circuital simulation environments provide a variety of custom circuit simulation tools to quickly and easily evaluate the designed circuit. In this second phase, the circuit, if well designed, will be checked to evaluate if it works as expected from the analysis carried on in the first phase of the procedure.

Third step
Finally, the circuit is physically implemented and experimentally characterized. The range of operating frequencies of each state variable is an important parameter of the system for both the feasibility and the ability to simultaneously acquire the trend of the state variables of the circuit. The acquisition of waveforms generated by the circuits, in fact, is a necessary practice that is performed in order to analyze in more detail the individual behavior of all the state variables and compare them with the theoretical trends that have been obtained in the numerical simulations.

2.3 An Example: The Rössler System

The Rössler system is one of the most well-known autonomous nonlinear systems that exhibit a chaotic attractor [5]. In terms of dimensionless equations the Rössler system is described by

$$\begin{cases} \dot{x} = -y - z \\ \dot{y} = x + ay \\ \dot{z} = b + z(x - c) \end{cases} \tag{2.32}$$

where

$$a = 0.2; \quad b = 0.2; \quad c = 7.0 \tag{2.33}$$

are parameter values for which a chaotic attractor appears (other values, however, also leading to chaotic behavior, are possible). In the following, the three steps of the methodology for the design of an electronic circuit equivalent to the Rössler system are applied to Eq. (2.32) to illustrate the procedure.

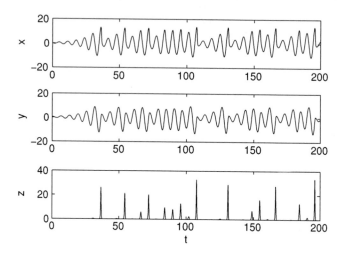

Fig. 2.16 Numerical simulations of Eq. (2.32): trend of the state variables x, y and z

Fig. 2.17 Numerical simulations of Eq. (2.32): chaotic attractor

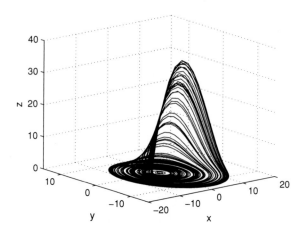

First step

Using *MATLAB*® and the mathematical model in Eq. (2.32) we analyze the temporal trends of the three state variables x, y, z, in particular paying attention to the amplitude range inside which each state variable oscillates. Figure 2.16 shows the behavior of the variables x, y, and z. In Fig. 2.17 the typical chaotic attractor of the Rössler system is shown. The plot is obtained by using the parameters (2.33). The ranges inside which x, y, and z oscillate, are the following:

$$x \in (-20, 20); \quad y \in (-15, 15); \quad z \in (0, 40) \qquad (2.34)$$

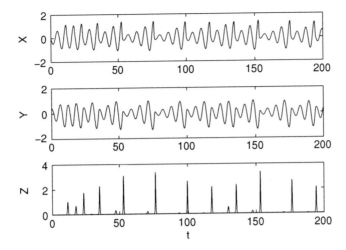

Fig. 2.18 Numerical simulations of Eq. (2.37): trend of the state variables X, Y and Z

If we use common 9 V batteries, these intervals are too large, so the system in Eq. (2.32) must be rescaled in amplitude. Design choices lead us to choose the scale factors as follows:

$$X = k_1 x; \qquad Y = k_2 y; \qquad Z = k_3 z \tag{2.35}$$

where

$$k_1 = \frac{1}{10}; \qquad k_2 = \frac{1}{10}; \qquad k_3 = \frac{1}{10}. \tag{2.36}$$

In this way a new rescaled equivalent system is found:

$$\begin{cases} \dot{X} = -Y - Z \\ \dot{Y} = X + aY \\ \dot{Z} = \frac{b}{10} + 10XZ - cZ. \end{cases} \tag{2.37}$$

The new rescaled system is now simulated and the trends of the state variables are verified to oscillate inside voltage limits that are now realizable. In this way, the feasibility of the circuit is checked (Figs. 2.18 and 2.19).

Second step
In this step, using operational amplifiers, resistors, capacitors, and other electronic components an equivalent electronic circuit is designed and then simulated with a circuital simulator to verify that the trends of the state variables are consistent with the numerical simulations analyzed before.

We start discussing the part of the circuit associated to the first rescaled equation

Fig. 2.19 Numerical
simulations of Eq. (2.37):
chaotic attractor

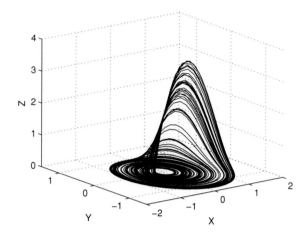

Fig. 2.20 Scheme of the
circuit associated to the first
rescaled Rössler equation.
Parameters are: $R_1 = 100\,\mathrm{k\Omega}$,
$R_2 = 100\,\mathrm{k\Omega}$, $R_3 = 100\,\mathrm{k\Omega}$,
$R_4 = 100\,\mathrm{k\Omega}$, $R_5 = 100\,\mathrm{k\Omega}$,
$R_6 = 1\,\mathrm{k\Omega}$, $R_7 = 33.3\,\mathrm{k\Omega}$,
$C_1 = 100\,\mathrm{nF}$, $V_{cc} = 9\,\mathrm{V}$

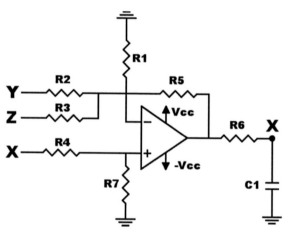

$$\dot{X} = -Y - Z. \tag{2.38}$$

The implementation of this equation needs one RC integrator and an adder. To the aim of using the RC integrator, the equation is rewritten in the form (2.15):

$$\dot{X} = -X + X - Y - Z. \tag{2.39}$$

Starting from this equation and keeping in mind the gain rule (2.10), we derive the circuit scheme of Fig. 2.20. Off-the-shelf resistors and capacitors, each with its own tolerance, are chosen as reported in the caption of Fig. 2.20. The choice of C_1 and R_6 fixes the time rescaling as $\kappa = \frac{1}{R_6 C_1} = 10000$. The circuit obeys the equation:

$$C_1 R_6 \dot{X} = -X + \frac{R_5}{R_4} X - \frac{R_5}{R_2} Y - \frac{R_5}{R_3} Z. \tag{2.40}$$

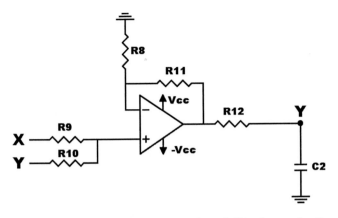

Fig. 2.21 Scheme of the circuit associated to the second rescaled Rössler equation. Parameters are:
$R_8 = 78\,k\Omega$, $R_9 = 100\,k\Omega$, $R_{10} = 78\,k\Omega$, $R_{11} = 100\,k\Omega$, $R_{12} = 1\,k\Omega$, $C_2 = 100\,nF$, $V_{cc} = 9\,V$

In a similar way, we proceed for the other two state variables. The second rescaled equation is rewritten as:

$$\dot{Y} = -Y + X + (a + 1)Y. \tag{2.41}$$

The scheme of the circuit is reported in Fig. 2.21 and the associated circuital equation as the following:

$$C_2 R_{12} \dot{Y} = -Y + \frac{R_{11}}{R_9}X + \frac{R_{11}}{R_{10}}Y. \tag{2.42}$$

Finally, the third rescaled equation is dealt with. The associated equation is rewritten as:

$$\dot{Z} = -Z + \frac{b}{10} + 10XZ + Z(1 - c). \tag{2.43}$$

In this equation two new types of terms appear, a constant term $\frac{b}{10}$ and a nonlinear term XY. To implement the first term a voltage divider is used. The second term is realized through the analog multiplier AD633, as shown in Fig. 2.22, taking into account that the output is given by:

$$W = \frac{(X_1 - X_2)(Y_1 - Y_2)}{10\,V} \frac{R_{22} + R_{23}}{R_{22}}. \tag{2.44}$$

So, if we select

$$X1 = X, \qquad X2 = 0, \qquad Y1 = Y, \qquad Y2 = 0 \tag{2.45}$$

and the ratio $\frac{R_{22}+R_{23}}{R_{22}} = 10$, we obtain the term XZ. In summary, the third circuit equation is the following:

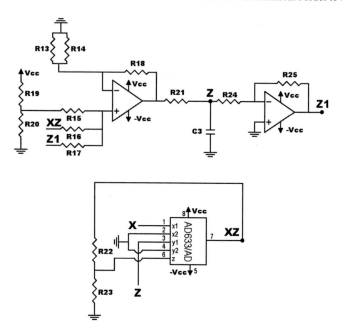

Fig. 2.22 Scheme of the circuit associated to the third rescaled Rössler equation. Parameters are: $R_{13} = 100\,k\Omega$, $R_{14} = 10\,k\Omega$, $R_{15} = 100\,k\Omega$, $R_{16} = 10\,k\Omega$, $R_{17} = 100\,k\Omega$, $R_{18} = 100\,k\Omega$, $R_{19} = 449\,k\Omega$, $R_{20} = 1\,k\Omega$, $R_{21} = 1\,k\Omega$, $R_{22} = 1\,k\Omega$, $R_{23} = 9\,k\Omega$, $R_{24} = 10\,k\Omega$, $R_{25} = 80\,k\Omega$ (potentiometer), $C_3 = 100\,nF$, $V_{cc} = 9\,V$

$$C_3 R_{21} \dot{Z} = -Z + \frac{R_{18} R_{25}}{R_{17} R_{24}} Z + \frac{R_{18}}{R_{16}} XZ + \frac{R_{18}}{R_{15}} \frac{R_{20}}{R_{19} + R_{20}} V_{cc} \qquad (2.46)$$

The whole circuit is obtained by assembling the three parts of Figs. 2.20, 2.21 and 2.22, so that the electronic circuit equivalent to the Rössler system is the circuit shown in Fig. 2.23. We notice that $C_1 R_6 = C_2 R_{12} = C_3 R_{21}$ so that the time scaling is coherent for all the equations of the set. Once designed the Rössler circuit scheme, it is checked through a circuital simulation tool, to verify that the trends of state variables are in agreement with the theoretical expectations.

Third step

In this final step, the Rössler circuit is physically implemented with low-cost components and welded on a predrilled board as shown in Fig. 2.24. The circuit is powered by a voltage generator and an oscilloscope is used to analyze the circuit behavior: the typical Rössler attractor shown in Fig. 2.25 is found. To compensate for all the component tolerances and to reproduce the different behaviors of the Rössler circuit it might be necessary to change the parameter c, which can be suitably accomplished by varying the value of the resistor R_{25}. A variable resistor, i.e., a trimmer, is then used substituting R_{25}.

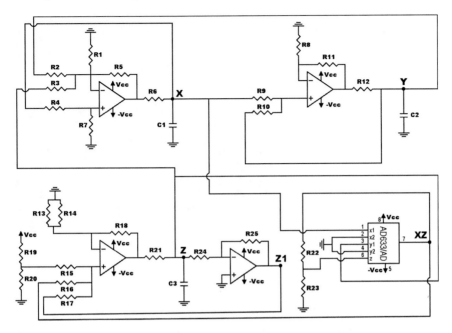

Fig. 2.23 Scheme of the electronic circuit equivalent to the Rössler system

Fig. 2.24 A picture of the implemented Rössler circuit

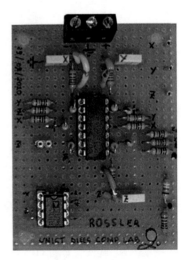

2.4 Implementation Through FPAA

Field Programmable Analog Arrays (FPAAs) provide an alternative way to implement chaotic circuits, following the same approach described in Sect. 2.2, but using, instead of components to be mounted on a development board, the blocks which are

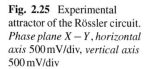

Fig. 2.25 Experimental attractor of the Rössler circuit. *Phase plane X − Y, horizontal axis* 500 mV/div, *vertical axis* 500 mV/div

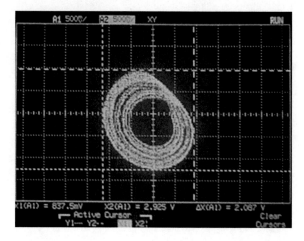

already contained in such programmable analog device. In fact, FPAAs represent the analog counterpart of Field Programmable Gate Array (FPGA) and contain a matrix of Configurable Analog Blocks (named CABs) that can be connected with each other and with external I/O blocks. Each CAB typically contains digital and analog comparators, some operational amplifiers and a series of capacitors. The FPAA technology is, in fact, mainly based on switched capacitor technology. The CAB blocks are surrounded by the other elements of the device, dedicated to clock management, signal I/O and block configuration, and dynamic reprogrammability. Some device can be also connected to external microcontroller to offer on-the-fly dynamic reprogrammability of the parameter values. Typical CABs are: inverting gain block, integrators, analog filters, algebraic adders. An FPAA in practice contains the fundamental blocks which are needed for an implementation based on the procedure of Sect. 2.2. The design can be done following the same guidelines, by taking into account the voltage supply limits which now depend on the specific hardware equipment used.

Using FPAA it is possible to reprogram the entire circuit dynamics, keeping the structure fixed but changing the parameters. The reprogrammability features of FPAA can also be used to adapt the circuit to changing external conditions due to noise or changes in the operating conditions of the system being controlled. The circuit configurations can be changed at a low level, where components such as operational amplifiers, capacitors, resistors, transconductors, and current mirrors can easily be fixed and connected, and also at a high level. In the latter case, user-friendly tools are often made available by the producers of the devices in order to reduce the time to market for products.

The two main characteristics of FPAA are the possibility to translate complex analog circuits into a set of low-level functions and the capability to place analog circuits under real-time software control within the system. For these reasons FPAA provides an interesting approach to implement chaotic circuits with programmable features. Examples of the use of FPAA for chaotic circuit implementations will be provided in Chap. 4.

References

1. Sedra AS, Smith KC (2003) Microelectronic circuits. Oxford University Press, Oxford
2. Fortuna L, Rizzo A, Xibilia MG (2003) Modeling complex dynamics via extended PWL-based CNNs. Int J Bifurcat Chaos 13(11):3273–3286
3. Xia Y, Fu M, Shi P (2009) Analysis and synthesis of dynamical systems with time-delays. Lecture notes in control and information sciences. Springer, New York
4. Ikeda K, Matsumoto K (1987) High-dimensional chaotic behavior in systems with time-delayed feedback. Physica D 29:223–235
5. Rossler OE (1976) An equation for continuous chaos. Phys Lett A 57(5):397–398

Chapter 3
A Gallery of Chaotic Circuits

Abstract In this chapter, a gallery of nonlinear chaotic circuits is presented. Each section deals with a specific circuit derived from the mathematical model of a nonlinear system. The electrical scheme and a sample of the behavior that the circuit can generate are reported, so that the reader can find a reference for his/her own experiments. Examples of both autonomous and nonautonomous circuits are presented.

Keywords Chaos · Chaotic circuits · Design of chaotic circuits

3.1 The Jerk Circuit

The first circuit dealt with in this chapter is an example of an autonomous circuit obtained from one member of the class of dynamical systems described by the nonlinear differential equation:

$$\dddot{x} = g(\ddot{x}, \dot{x}, x) \tag{3.1}$$

which is referred to as the *jerk equation*. In particular, we consider the system [1]

$$\dddot{x} + A\ddot{x} + x + f(\dot{x}) = 0 \tag{3.2}$$

where the nonlinear function $f(\dot{x})$ is given by $f(\dot{x}) = IR(e^{\frac{\dot{x}}{\alpha}} - 1)$. For $I = 10^{-12}$, $R = 10^3$, $\alpha = 0.026$, and $A = 1$ the solution of Eq. (3.2) is a chaotic trajectory. Once fixed the nonlinear function $f(\dot{x})$, the nonlinear differential equation (3.2) is rewritten in state-space form, hence as a set of three differential equations:

A. Buscarino et al., *A Concise Guide to Chaotic Electronic Circuits*,
SpringerBriefs in Applied Sciences and Technology,
DOI: 10.1007/978-3-319-05900-6_3, © The Author(s) 2014

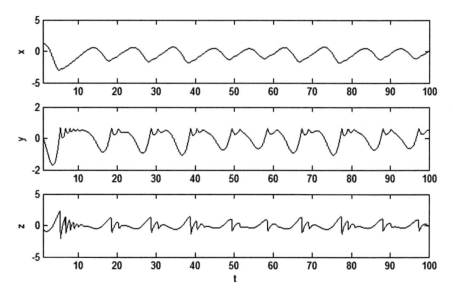

Fig. 3.1 Numerical simulations of Eq. (3.3): trend of the state variables x, y, and z

Fig. 3.2 Numerical
simulations of Eq. (3.3):
chaotic attractor

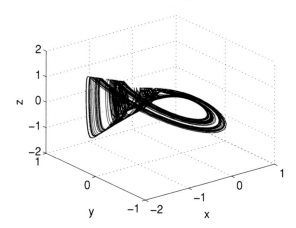

$$\begin{cases} \dot{x} = y \\ \dot{y} = z \\ \dot{z} = -z - Ax - 10^{-9}(e^{\frac{y}{\alpha}} - 1) \end{cases} \tag{3.3}$$

The first step for the electronic implementation of Eq. (3.3) is the study of the range where the oscillations of the individual state variables are confined. To this aim, the model is simulated with *MATLAB*®. The trend of the three-state variables x, y, and z and the corresponding attractor are shown in Figs. 3.1 and 3.2.

Fig. 3.3 Circuit
implementation of the jerk
dynamics (3.3). Components:
D = 1N4148 Diode,
R_1 = 1 kΩ, R_2 = 1 kΩ,
R_3 = 1 kΩ, R_4 = 1 kΩ,
R_5 = 1 kΩ, R_6 = 1 kΩ,
C_1 = 1 μF, C_2 = 1 μF,
C_3 = 1 μF, Operational
Amplifier = TL084, V_{cc} = 9 V

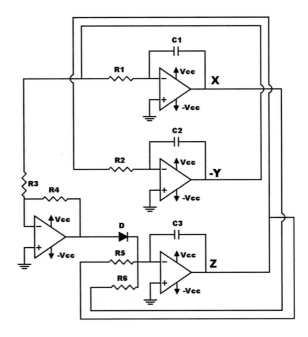

As it can be noticed, all the three variables oscillate in a range which is compatible
with a voltage power supply fixed to ± 9 V, so the variables do not need to be
scaled. The second step of the procedure involves the design and circuit simulation
of the electronic system. The nonlinear system in Eq. (3.3) has been implemented
with an electronic circuit using three basic blocks based on the Miller integrator
configuration. In Fig. 3.3 the circuit implementing the jerk dynamics (3.3) is shown.
It obeys to the following circuital equations:

$$\begin{cases} C_1 R_1 \frac{dX}{d\tau} = y \\ C_2 R_2 \frac{dY}{d\tau} = z \\ C_3 R_5 \frac{dZ}{d\tau} = -z - \frac{R_5}{R_6} x - R_5 h \end{cases} \quad (3.4)$$

where $h = 10^{-9}(e^{\frac{y}{\alpha}} - 1)$ is implemented using a diode and a temporal rescaling
$\kappa = \frac{1}{C_1 R_1} = \frac{1}{C_2 R_2} = \frac{1}{C_3 R_5} = 1000$. Circuit simulations confirm that the expected
waveforms for the state variables are retrieved.

The third step deals with the experimental realization of the electrical scheme of
Fig. 3.3. This requires a small number of off-the-shelf components, which can be
mounted on a small board as shown in Fig. 3.4.

In Fig. 3.5, the chaotic attractor of the circuit, as obtained in the oscilloscope (plane
$x-y$), is reported. The trend of the state variables x and y are shown in Fig. 3.6.

In Fig. 3.7, an experimental bifurcation diagram is shown. The diagram is obtained
by performing a set of acquisitions of the circuit waveforms at different values of

Fig. 3.4 A picture of the
circuit implementing the jerk
dynamics (3.2)

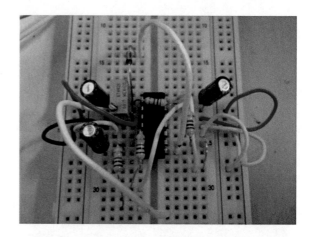

Fig. 3.5 Experimental results
of the circuit implementing
the jerk dynamics (3.2):
projection of the attractor
in the phase plane x–
y. Horizontal axis =
500 mV/div, vertical axis =
500 mV/div

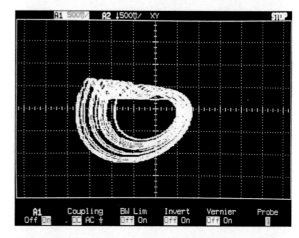

Fig. 3.6 Experimental results
of the circuit implementing
the jerk dynamics (3.2):
waveforms of the state
variables x and y

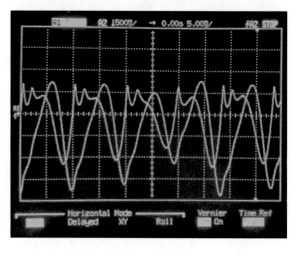

Fig. 3.7 Experimental bifurcation diagram for the circuit implementing the jerk dynamics (3.2)

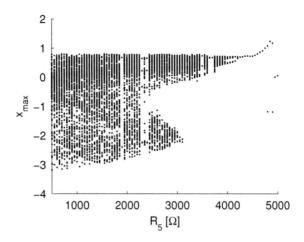

the resistor R_6, actually a potentiometer, implementing parameter $A = \frac{R_5}{R_6}$. For each value of R_6, the local maxima of state variable x are reported: when the circuit has a periodic behavior, the number of local maxima represents the periodicity of the limit cycle, while chaos is characterized by a continuum of points. As expected from numerical simulations, different windows of chaotic behavior can be found ranging A from 0.2 to 2.

3.2 The Chua's Circuit

As mentioned in Chap. 1, the Chua's circuit is the simplest autonomous third-order nonlinear electronic circuit with a rich variety of dynamical behaviors including chaos. It is in fact considered the canonical circuit for studying chaos. The dimensionless equations of the Chua's circuit can be derived from Eq. (1.1) and read as follows:

$$\begin{cases} \dot{x} = \alpha(y - h(x)) \\ \dot{y} = x - y + z \\ \dot{z} = -\beta y \end{cases} \tag{3.5}$$

where $h(x)$ represents the nonlinearity of the system:

$$h(x) = m_1 + \frac{1}{2}(m_0 - m_1)(|x + 1| - |x - 1|) \tag{3.6}$$

In this section starting from Eq. (3.5), we will discuss an implementation of the Chua's circuit based on operational amplifiers. In this implementation, the three state variables represent physical variables different from that of the original circuit. In particular, being all associated to voltages across capacitors (according to the guidelines of the approach which makes use of RC integrators), the new circuit makes

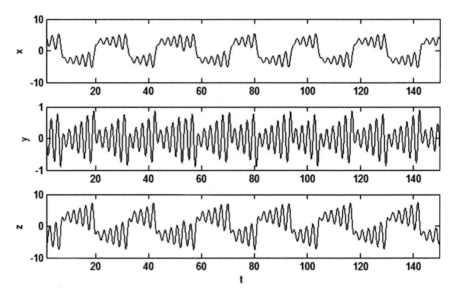

Fig. 3.8 Numerical simulations of Eq. (3.5): trend of the state variables x, y, and z

no use of the inductor. Furthermore, this circuit is directly mapped into a Cellular Nonlinear Network [2] and so, it may be realized in this framework. On the other hand, the circuit is perfectly equivalent to the original one.

The double-scroll chaotic attractor is obtained in Eq. (3.5) for the following values of the parameters

$$m_0 = -\frac{1}{7} \qquad m_1 = \frac{2}{7} \qquad \alpha = 9 \qquad \beta = 14.286 \tag{3.7}$$

In fact, numerical simulations of Eq. (3.5) obtained with *MATLAB®* and reported in Figs. 3.8 and 3.9, confirm the appearance of the double-scroll chaotic attractor. Other chaotic attractors as well as many other nonlinear phenomena are obtained for other sets of the circuit parameters.

Following the methodology described in Chap. 2, an electronic circuit has been designed as reported in Fig. 3.10 and implemented as shown in Fig. 3.11 realized using resistors, capacitors, and operational amplifiers. We notice that, following the approach of Chap. 2, one of the possible implementations of the Chua's circuit dynamics is obtained. Several other implementations, including the original formulation of the Chua's circuit, which was done in terms of one inductor, two capacitors, one linear resistor, and one nonlinear resistor, realized through operational amplifiers, are discussed in [3].

The circuital equations associated to the implementation of Chua's circuit are the following:

Fig. 3.9 Numerical simulations of Eq. (3.5): chaotic attractor

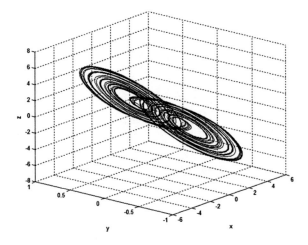

$$\begin{cases} C_1 R_6 \frac{\mathrm{d}X}{\mathrm{d}\tau} = -X + \frac{R_5}{R_3}Y + \frac{R_5}{R_2}h \\ C_2 R_{18} \frac{\mathrm{d}Y}{\mathrm{d}\tau} = -Y + \frac{R_{17}}{R_{14}}X + \frac{R_{17}}{R_{15}}Z \\ C_3 R_{23} \frac{\mathrm{d}Z}{\mathrm{d}\tau} = -Z + \frac{R_{21}}{R_{20}}Z + \frac{R_{21}}{R_{19}}Y \end{cases} \tag{3.8}$$

where:

$$h = \frac{R_{12}}{R_{11}+R_{12}} \frac{R_9}{R_8}(|X + 1| - |X - 1|) \tag{3.9}$$

Matching Eq. (3.8) with the mathematical model in Eq. (3.5) leads to choose commercial values of resistors, introducing a temporal rescaling $\kappa = \frac{1}{C_2 R_{18}} = \frac{1}{C_3 R_{23}} = 10000$. The different dynamical behaviors shown by the Chua's circuit by varying the single bifurcation parameter α can be observed in the circuit by varying resistor R_6, according to the relation $\alpha = \frac{R_5}{R_3} \frac{R_{18}}{R_6}$. In Figs. 3.12 and 3.13, a typical chaotic behavior and a limit cycle both observed in the experimental circuit are shown.

3.3 The Lorenz System

The Lorenz system [4] is a system of ordinary differential equations first studied by Edward Lorenz. It has been developed in order to obtain a simplified model for the atmospheric convection. It consists of the following three differential equations:

$$\begin{cases} \dot{x} = \alpha(y - x) \\ \dot{y} = \rho x - xz - y \\ \dot{z} = xy - \beta z \end{cases} \tag{3.10}$$

where the parameter values can be chosen as:

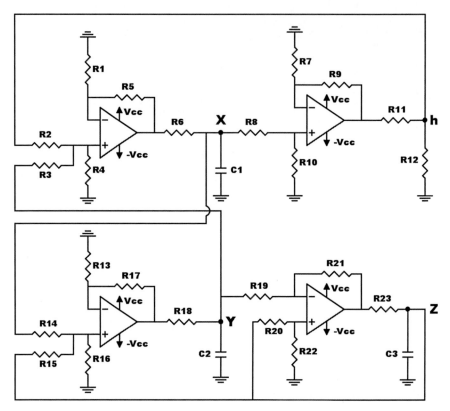

Fig. 3.10 Circuit implementation of Eq. (3.5). Components: $R_1 = 4$ kΩ, $R_2 = 13.3$ kΩ, $R_3 = 5.6$ kΩ, $R_4 = 20$ kΩ, $R_5 = 20$ kΩ, $R_6 = 380$ Ω (potentiometer), $R_7 = 112$ kΩ, $R_8 = 112$ kΩ, $R_9 = 1$ MΩ, $R_{10} = 1$ MΩ, $R_{11} = 12.1$ kΩ, $R_{12} = 1$ kΩ, $R_{13} = 51.1$ kΩ, $R_{14} = 100$ kΩ, $R_{15} = 100$ kΩ, $R_{16} = 100$ kΩ, $R_{17} = 100$ kΩ, $R_{18} = 1$ kΩ, $R_{19} = 8.2$ kΩ, $R_{20} = 100$ kΩ, $R_{21} = 100$ kΩ, $R_{22} = 7.8$ kΩ, $R_{23} = 1$ kΩ, $C_1 = C_2 = C_3 = 100$ nF, $V_{cc} = 9$ V

Fig. 3.11 A picture of the
Chua's circuit

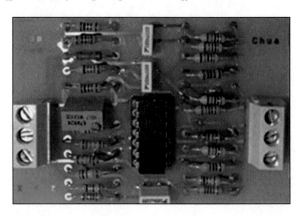

Fig. 3.12 Experimental attractor observed from the Chua's circuit in the phase plane x–y. Horizontal axis = 500 mV/div, vertical axis = 100 mV/div

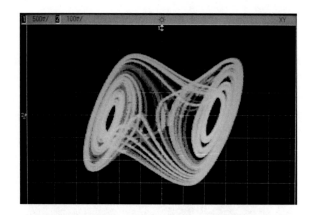

Fig. 3.13 Experimental limit cycle observed from the Chua's circuit in the phase plane x–y. Horizontal axis = 500 mV/div, vertical axis = 100 mV/div

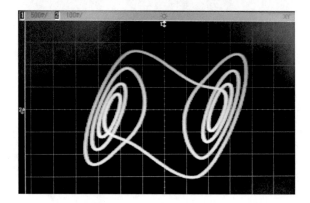

$$\alpha = 10, \rho = 28, \beta = \frac{8}{3} \tag{3.11}$$

in order to obtain a chaotic behavior.

Simulating model (3.10) with *MATLAB*®, the temporal evolution of the three state variables x, y, and z can be observed. The numerical simulation shown in Fig. 3.14 allows to identify the oscillation ranges as follows:

$$x \in (-20, 20); \quad y \in (-20, 20); \quad z \in (0, 50) \tag{3.12}$$

In Fig. 3.15, using the parameters in (3.11), a typical chaotic attractor for the Lorenz circuit is shown. If common 9 V batteries are used, the intervals (3.12) cannot be realized, so the system in Eq. (3.10) must be rescaled in amplitude. These considerations lead us to choose the scale factors as follows:

$$X = k_1 x; \quad Y = k_2 y; \quad Z = k_3 z. \tag{3.13}$$

where

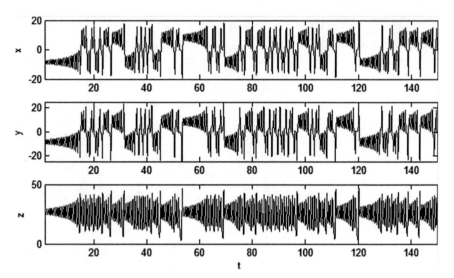

Fig. 3.14 Numerical simulations of Eq. (3.10): trend of the state variables x, y, and z

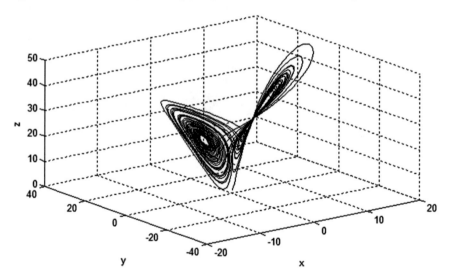

Fig. 3.15 Numerical simulations of Eq. (3.10): chaotic attractor

$$k_1 = \frac{1}{10}; \quad k_2 = \frac{1}{10}; \quad k_3 = \frac{1}{30}. \tag{3.14}$$

In this way, a new rescaled set of variables is obtained:

$$x = \frac{X}{k_1}; \quad y = \frac{Y}{k_2}; \quad z = \frac{Z}{k_3}. \tag{3.15}$$

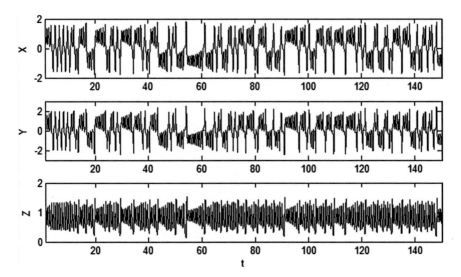

Fig. 3.16 Numerical simulations of Eq. (3.16): trend of the state variables X, Y, and Z

Replacing x, y, and z of Eq. (3.15) in system (3.10) the new rescaled equivalent system is:

$$\begin{cases} \dot{X} = \alpha(Y - X) \\ \dot{Y} = \rho X - 30XZ - Y \\ \dot{Z} = \frac{100XY}{30} - \beta Z \end{cases} \tag{3.16}$$

The new rescaled state variables oscillate inside realizable voltage limits (Figs. 3.16, 3.17).

Now the methodology described in Chap. 2 is applied to design the circuit shown in Fig. 3.18.

The circuital equations associated to the implementation of Lorenz's circuit are the following:

$$\begin{cases} C_1 R_5 \frac{dX}{d\tau} = -X - \frac{R_4}{R_1}X + \frac{R_4}{R_2}X + \frac{R_4}{R_3}Y \\ C_2 R_{11} \frac{dY}{d\tau} = -Y - \frac{R_{10}}{R_7}XZ + \frac{R_{10}}{R_8}X \\ C_3 R_{17} \frac{dZ}{d\tau} = -Z - \frac{R_{16}}{R_{13}}Z + \frac{R_{16}}{R_{14}}XY \end{cases} \tag{3.17}$$

Matching equation (3.17) with the mathematical model in Eq. (3.16), the values of circuital components can be selected, fixing a time scaling $\kappa = \frac{1}{C_1 R_5} = \frac{1}{C_2 R_{11}} = \frac{1}{C_3 R_{17}} = 5000$. In Fig. 3.19 an hardware implementation of the Lorenz circuit and in Figs. 3.20 and 3.21 the typical chaotic behavior, known as the butterfly attractor, of the experimental circuit are shown.

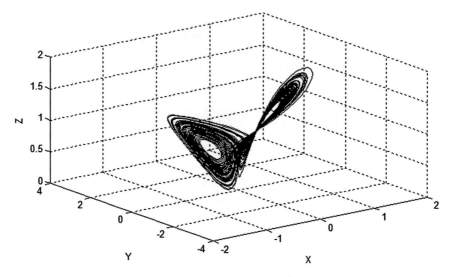

Fig. 3.17 Numerical simulations of Eq. (3.16): chaotic attractor

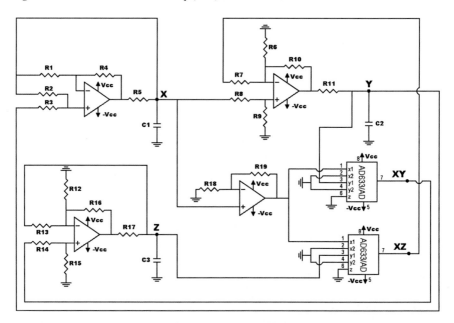

Fig. 3.18 Circuital implementation of Eq. (3.16). Components: $R_1 = 10\,\text{k}\Omega$, $R_2 = 100\,\text{k}\Omega$, $R_3 = 10\,\text{k}\Omega$, $R_4 = 100\,\text{k}\Omega$, $R_5 = 1\,\text{k}\Omega$, $R_6 = 5.6\,\text{k}\Omega$, $R_7 = 3.3\,\text{k}\Omega$, $R_8 = 3.6\,\text{k}\Omega$, $R_9 = 3.19\,\text{k}\Omega$, $R_{10} = 100\,\text{k}\Omega$, $R_{11} = 1\,\text{k}\Omega$, $R_{12} = 3.3\,\text{k}\Omega$, $R_{13} = 37.5\,\text{k}\Omega$, $R_{14} = 3.3\,\text{k}\Omega$, $R_{15} = 3.74\,\text{k}\Omega$, $R_{16} = 100\,\text{k}\Omega$, $R_{17} = 1\,\text{k}\Omega$, $R_{18} = 1\,\text{k}\Omega$, $R_{19} = 9\,\text{k}\Omega$, $C_1 = 200\,\text{nF}$, $C_2 = 200\,\text{nF}$, $C_3 = 200\,\text{nF}$, $V_{cc} = 9\,\text{V}$

Fig. 3.19 Picture of
the implemented circuit
reproducing the Lorenz
system

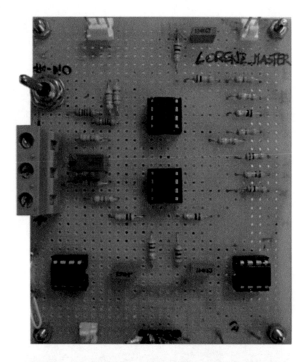

Fig. 3.20 Experimental
results of the Lorenz
circuit: chaotic attractor
shown by the implemented
circuit. Phase plane:
X–Y. Horizontal axis $=$
200 mV/div, vertical axis $=$
500 mV/div

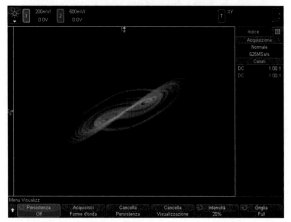

3.4 The Hindmarsh–Rose Neuron

In traditional artificial neural networks, the neuron behavior is described only in
terms of firing rate, while most real neurons, commonly known as spiking/bursting
neurons, transmit information by pulses or bursts of pulses, also called action poten-
tials or spikes. The Hindmarsh-Rose (HR) model [5] is computationally simple and

Fig. 3.21 Experimental results of the Lorenz circuit: chaotic attractor shown by the implemented circuit. Phase plane: X–Z. Horizontal axis = 200 mV/div, vertical axis = 500 mV/div

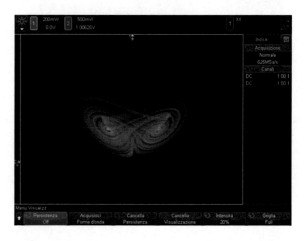

capable of reproducing rich firing patterns exhibited by real biological neurons. It consists of three coupled ordinary differential equations that, as a function of the parameter values, can generate different spiking and bursting behaviors, including chaotic spiking.

The Hindmarsh-Rose neuron is described by the following dimensionless equations:

$$\begin{cases} \dot{x} = y - ax^3 + bx^2 - z + I \\ \dot{y} = c - dx^2 - y \\ \dot{z} = r(s(x - \chi) - z) \end{cases} \tag{3.18}$$

where x represents the membrane potential, y and z are associated with fast and slow current channels, respectively, and the parameters of the system when a chaotic bursting can be observed are:

$$a = 1, b = 3, c = 1, d = 5, I = 3.281, \chi = -1.6, r = 0.002, s = 4 \tag{3.19}$$

The numerical simulation shown in Figs. 3.22 and 3.23 allows to observe the dynamics of the three state variables x, y, and z. In the specific case, a chaotic bursting is obtained for parameter values as in Eq. (3.19).

Although the three state variables oscillate within the supply voltage range, combinations of them involved in the state equations are outside the feasible range. Hence, it is necessary to apply a rescaling as reported below:

$$X = k_1 x; \qquad Y = k_2 y; \qquad Z = k_3 z \tag{3.20}$$

where

$$k_1 = \frac{1}{2.5}; \qquad k_2 = \frac{1}{2}; \qquad k_3 = 1. \tag{3.21}$$

The rescaled equivalent system is governed by the following equations:

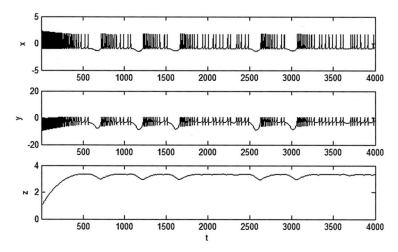

Fig. 3.22 Numerical simulations of Eq. (3.18): trend of the state variables x, y, and z

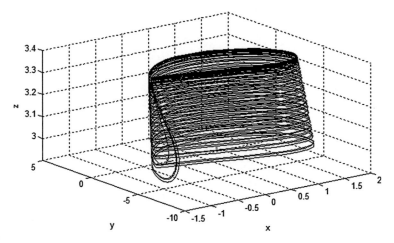

Fig. 3.23 Numerical simulations of Eq. (3.18): chaotic attractor

$$\begin{cases} \dot{X} = 3.33Y - 9aX^3 + 3bX^2 - \frac{2}{3}Z + \frac{I}{3} \\ \dot{Y} = \frac{c}{10} - 0.9dX^2 - Y \\ \dot{Z} = \frac{r}{2}(s(3X - \chi) - 2Z) \end{cases} . \tag{3.22}$$

The numerical simulation of Eq. (3.22) shown in Figs. 3.24 and 3.25 confirms that these equations can be implemented without saturations. The electronic circuit realized using resistors, capacitors, diodes, and operational amplifiers mimicking Eq. (3.22) is shown in Fig. 3.26. In the design of Hindmarsh-Rose circuit, for the implementation of the circuitry performing the cube and the square functions, the PWL-based circuital schemes shown in Chap. 2, are used.

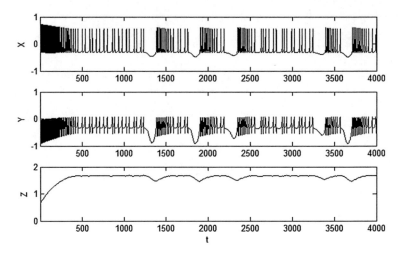

Fig. 3.24 Numerical simulations of Eq. (3.22): trend of the state variables X, Y, and Z

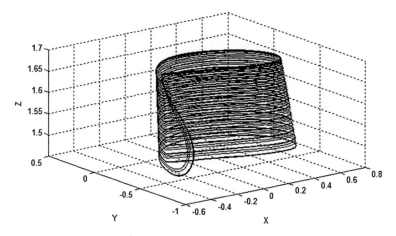

Fig. 3.25 Numerical simulations of Eq. (3.22): chaotic attractor

The circuital equations associated to the Hindmarsh-Rose circuit are the following:

$$
\begin{cases}
C_1 R_{11} \frac{dX}{d\tau} = -X - \frac{R_1}{R_3}Z + \frac{R_1}{R_4}X + \frac{R_1}{R_5}X^2 - \frac{R_1}{R_6}X^3 + \frac{R_1}{R_7}Y + \frac{R_{10}}{R_9+R_{10}}\frac{R_1}{R_8}V_{cc} \\
C_2 R_{19} \frac{dY}{d\tau} = -Y - \frac{R_{12}}{R_{13}}X^2 + \frac{R_{12}}{R_{14}}Y + \frac{R_{12}}{R_{17}}\frac{R_{16}}{R_{15+16}}V_{cc} \\
C_3 R_{27} \frac{dZ}{d\tau} = -Z + \frac{R_{20}}{R_{22}}X + \frac{R_{20}}{R_{23}}\frac{R_{25}}{R_{24}+R_{25}}V_{cc}
\end{cases}
$$

(3.23)

The parameters of the HR model, i.e. b, c, d, and I, are related to the circuit components through the following relationships:

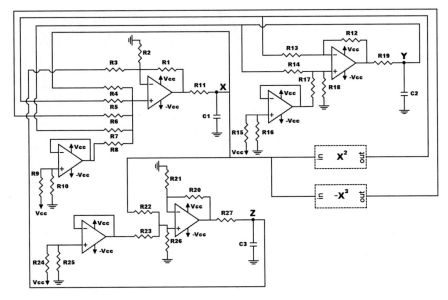

Fig. 3.26 Circuital implementation of Hindmarsh-Rose neuron. Components: $R_1 = 200$ kΩ, $R_2 = 10$ kΩ, $R_3 = 500$ kΩ, $R_4 = 200$ kΩ, $R_5 = 26.6$ kΩ, $R_6 = 32$ kΩ, $R_7 = 250$ kΩ, $R_8 = 500$ kΩ, $R_9 = 2.17$ kΩ, $R_{10} = 1$ kΩ, $R_{11} = 1$ kΩ, $R_{12} = 100$ kΩ, $R_{13} = 32$ kΩ, $R_{14} = 125$ kΩ, $R_{15} = 89$ kΩ, $R_{16} = 1$ kΩ, $R_{17} = 100$ kΩ, $R_{18} = 43$ kΩ, $R_{19} = 200\Omega$, $R_{20} = 200$ kΩ, $R_{21} = 10$ kΩ, $R_{22} = 20$ kΩ, $R_{23} = 100$ kΩ, $R_{24} = 1.81$ kΩ, $R_{25} = 1$ kΩ, $R_{26} = 22$ kΩ, $R_{27} = 1$ kΩ, $C_1 = 1$ μF, $C_2 = 1$ μF, $C_3 = 467$ μF, $Vcc = 9$ V

$$b = \frac{1}{3}\frac{R_1}{R_5} \tag{3.24}$$

$$c = 10\frac{R_{16}}{R_{15}+16}\frac{R_{12}}{R_{17}}V_{cc} \tag{3.25}$$

$$d = \frac{1}{3}\frac{R_{12}}{R_{13}} \tag{3.26}$$

$$I = \frac{5}{2}\frac{R_{10}}{R_9+10}\frac{R_1}{R_8}V_{cc} \tag{3.27}$$

Furthermore, a time scale is introduced choosing $\kappa = \frac{1}{C_1 R_{11}} = 1000$. The bifurcation parameter I drives the system into different regions of chaos, limit cycle, and unstable region. To change the value of I, it is possible to use in the circuit shown in Fig. 3.26 a potentiometer replacing resistor R_9. Figure 3.27 reports the typical spikes observed on the oscilloscope from the real circuit.

Fig. 3.27 Experimental
results: bursting behavior
in the state variable x
fixing $R_9 = 127.16$.
Vertical axis $= 500$ mV/div

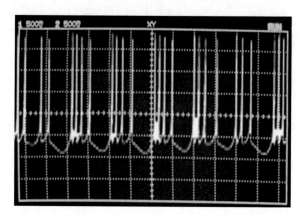

3.5 The Langford System

The mathematical model of the Langford system [6] is derived from the equations
of Navier-Stokes, and it has been used to describe the motion of turbulent flow in a
fluid:

$$\begin{cases} \dot{x} = xz - \omega y \\ \dot{y} = \omega x + xy \\ \dot{z} = p + z - \frac{1}{3}z^3 - (x^2 + y^2)(1 + qx + \varepsilon x) \end{cases} \tag{3.28}$$

where typical values for the system parameters are:

$$\omega = 10, p = 1.1, q = 0.7, \varepsilon = 0.5 \tag{3.29}$$

In Figs. 3.28 and 3.29, the behavior of the system (3.28) for parameters as in
(3.29) is shown. The three variables x, y, and z reach a maximum amplitude of 2,
but it is necessary to note that in the first two equations, the two state variables x
and y are multiplied by a parameter $\omega = 10$. As a result, the first two equations will
undergo two saturations introducing distortions in the behavior of the state variables.
To solve the problem, also in this case, a rescaling is required:

$$X = k_1 x; \qquad Y = k_2 y; \qquad Z = k_3 z \tag{3.30}$$

where

$$k_1 = \frac{1}{3}; \qquad k_2 = \frac{1}{3}; \qquad k_3 = \frac{1}{2}. \tag{3.31}$$

In this way, a new rescaled equivalent system, whose behavior is shown in Figs.
3.30 and 3.31, is obtained:

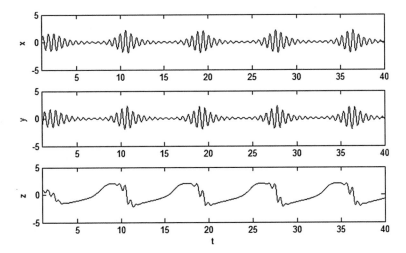

Fig. 3.28 Numerical simulations of Eq. (3.28): trend of the state variables x, y, and z

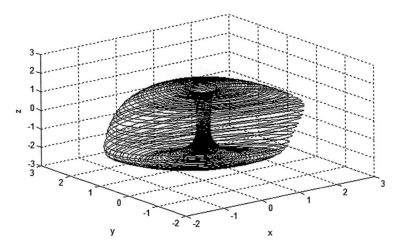

Fig. 3.29 Numerical simulations of Eq. (3.28): chaotic attractor

$$\begin{cases} \dot{X} = 2XZ - \omega Y \\ \dot{Y} = \omega X + 3XY \\ \dot{Z} = \frac{p}{2} + Z - \frac{4}{3}Z^3 - \frac{9}{2}(X^2 + Y^2)(1 + 3qX + 3\varepsilon X) \end{cases} \qquad (3.32)$$

The electronic circuit designed following the same adopted guidelines is reported in Fig. 3.32, in which a temporal rescaling with $\kappa = \frac{1}{C_1 R_{11}} = \frac{1}{C_2 R_{22}} = \frac{1}{C_3 R_{43}} = 20000$ has been introduced. In the design of the Langford circuit, for the implementation of the square function, it has been chosen the circuital scheme already described in Fig. 2.10. The cross-products are implemented using the AD633 whose output is amplified using a voltage divider.

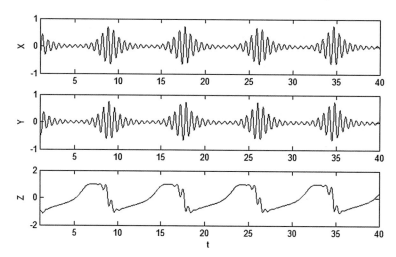

Fig. 3.30 Numerical simulations of Eq. (3.32): trend of the state variables X, Y, and Z

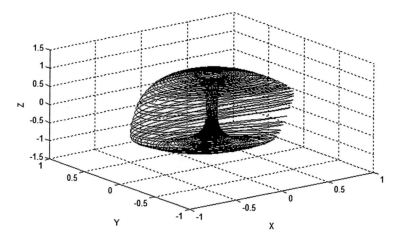

Fig. 3.31 Numerical simulations of Eq. (3.32): chaotic attractor

The circuital equations associated to the implementation of the Langford circuit are:

$$
\begin{cases}
C_1 R_{11} \frac{dX}{d\tau} = -X - \frac{R_1}{R_5}\frac{R_4}{R_3}Y + \frac{R_1}{R_8}XZ + \frac{R_1}{R_9}X \\
C_2 R_{22} \frac{dY}{d\tau} = -Y + \frac{R_{12}}{R_{16}}\frac{R_{15}}{R_{14}}X + \frac{R_{12}}{R_{19}}XY + \frac{R_{12}}{R_{20}}Y \\
C_3 R_{43} \frac{dZ}{d\tau} = -Z - \frac{R_{23}}{R_{26}}Z^3 \\
\qquad - \frac{R_{23}}{R_{37}}\left[\left(-\frac{R_{28}}{R_{27}}X^2 - \frac{R_{28}}{R_{29}}Y^2\right)\left(-\frac{R_{30}}{R_{31}}X - V_{cc}\frac{R_{30}}{R_{32}}\frac{R_{34}}{R_{33}+R_{34}}\right)\right] \\
\qquad + \frac{R_{23}}{R_{40}}\frac{R_{39}}{R_{38}+R_{39}}V_{cc} + \frac{R_{23}}{R_{41}}Z
\end{cases}
\tag{3.33}
$$

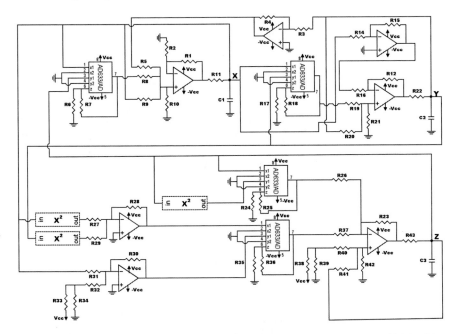

Fig. 3.32 Circuital implementation of the Langford model. Components: $R_1 = 100\,\text{k}\Omega$, $R_2 = 33.33\,\text{k}\Omega$, $R_3 = 10\,\text{k}\Omega$, $R_4 = 100\,\text{k}\Omega$, $R_5 = 100\,\text{k}\Omega$, $R_6 = 9\,\text{k}\Omega$, $R_7 = 1\,\text{k}\Omega$, $R_8 = 50\,\text{k}\Omega$, $R_9 = 100\,\text{k}\Omega$, $R_{10} = 100\,\text{k}\Omega$, $R_{11} = 5\,\text{k}\Omega$, $R_{12} = 100\,\text{k}\Omega$, $R_{14} = 100\,\text{k}\Omega$, $R_{15} = 1000\,\text{k}\Omega$, $R_{16} = 100\,\text{k}\Omega$, $R_{17} = 9\,\text{k}\Omega$, $R_{18} = 1\,\text{k}\Omega$, $R_{19} = 33.3\,\text{k}\Omega$, $R_{20} = 100\,\text{k}\Omega$, $R_{21} = 50\,\text{k}\Omega$, $R_{22} = 5\,\text{k}\Omega$, $R_{23} = 100\,\text{k}\Omega$, $R_{24} = 9\,\text{k}\Omega$, $R_{25} = 1\,\text{k}\Omega$, $R_{26} = 75\,\text{k}\Omega$, $R_{27} = 30\,\text{k}\Omega$, $R_{28} = 133\,\text{k}\Omega$, $R_{29} = 30\,\text{k}\Omega$, $R_{30} = 350\,\text{k}\Omega$, $R_{31} = 100\,\text{k}\Omega$, $R_{32} = 350\,\text{k}\Omega$, $R_{33} = 20\,\text{k}\Omega$, $R_{34} = 1.8\,\text{k}\Omega$, $R_{35} = 9\,\text{k}\Omega$, $R_{36} = 1\,\text{k}\Omega$, $R_{37} = 100\,\text{k}\Omega$, $R_{38} = 226\,\text{k}\Omega$, $R_{39} = 10\,\text{k}\Omega$, $R_{40} = 100\,\text{k}\Omega$, $R_{41} = 49.7\,\text{k}\Omega$, $R_{42} = 119.7\,\text{k}\Omega$, $R_{43} = 5\,\text{k}\Omega$, $C_1 = 10\,\text{nF}$, $C_2 = 10\,\text{nF}$, $C_3 = 10\,\text{nF}$, $V_{cc} = 12\,\text{V}$

The bifurcation parameter p in terms of the circuit components is set by:

$$p = \frac{R_{23}}{R_{40}} \frac{R_{39}}{R_{38} + R_{39}} V_{cc} \tag{3.34}$$

This bifurcation parameter p drives the system into different regions of nonlinear behavior. To change the value of p, it is possible to use in the circuit shown in Fig. 3.32 a potentiometer replacing resistor R_{38}. In Fig. 3.33 a realization of the circuit is reported, while in Fig. 3.34 the typical attractor of the Langford model observed on the oscilloscope is reported.

3.6 The Memristive Circuit

The memristor is the fourth basic circuit element, that can be defined as a dynamical resistor in which the resistance $R(w)$ is a function of the internal state variable w or in which there is a relationship between charge and magnetic flux linkage [7]. Due to

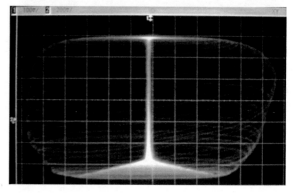

its intrinsic nonlinear characteristics, the memristor is a very interesting component
for the design of new dynamical circuits able to show complex behavior, like chaos.

In this section, a memristive chaotic circuit [8] is presented. The approach relies
on the design of a circuit equivalent to the memristor so that it can be realized with
common off-the-shelf components with the aim of observing the onset of new chaotic
attractors in nonlinear circuits with memristors. The memristive circuit is described
by the following set of equations:

$$\begin{cases} \dot{x} = \alpha(y - xH(w)) \\ \dot{y} = z - x \\ \dot{z} = -\beta y + \gamma z \\ \dot{w} = x \end{cases} \tag{3.35}$$

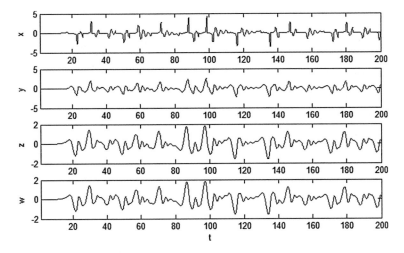

Fig. 3.35 Numerical simulations of Eq. (3.35): trend of the state variables x, y, z, and w

where

$$H(w) = \begin{cases} 0.2 \; |w| < 1 \\ 10 \; |w| > 1 \end{cases}$$

with $\alpha = 4$, $\beta = 1$, and $\gamma = 0.65$. As shown in the simulation reported in Figs. 3.35 and 3.36, the behavior of x, y, z, and w exhibits the typical features of a chaotic signal.

Despite the waveforms of the state variables x, y, z, and w have acceptable operating ranges, feasibility problems, due to the mathematical operations involving them and leading to terms over the voltage supply bounds as evidenced by simulations with the electronic circuit simulator *PSPICE*® appear. Thus, the variables of the circuit (in particular, x and w) are scaled with the scaling factors listed below:

$$X = k_1 x; \qquad Y = k_2 y; \qquad Z = k_3 z; \qquad W = k_4 w \qquad (3.36)$$

where

$$k_1 = \frac{1}{5}; \qquad k_2 = 1; \qquad k_3 = 1; \qquad k_4 = 2. \qquad (3.37)$$

In this way, a new rescaled equivalent system is obtained:

$$\begin{cases} \dot{X} = 0.8Y - 4XH(W) \\ \dot{Y} = Z - 5X \\ \dot{Z} = -Y + 0.65Z \\ \dot{W} = 10X \end{cases} \qquad (3.38)$$

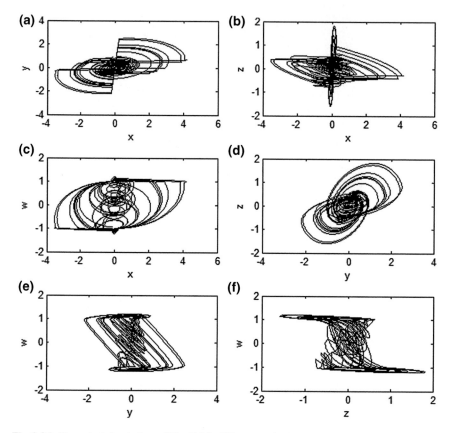

Fig. 3.36 Numerical simulations of Eq. (3.35): different projections of the chaotic attractor, **a** plane $x-y$; **b** plane $x-z$; **c** plane $x-w$; **d** plane $y-z$; **e** plane $y-w$; **f** plane $z-w$

with

$$H(w) = \begin{cases} 0.2 \; |w| < 2 \\ 10 \; |w| > 2 \end{cases}$$

Numerical simulations of the rescaled system are reported in Figs. 3.37 and 3.38. Before designing the electronic circuit of the system, some considerations on the components and on the particular configuration used to implement the memristive function are made. Let us consider the electronic scheme reported in Fig. 3.39, which is used for the implementation of the term $XH(W)$ in the memristive circuit. In order to demonstrate the operations performed by each stage, a sinusoidal input is provided as test signal. The first stage, implemented through an OP-AMP and a diode, is a full-wave rectifier performing the absolute value of the input. The next stage is an open-loop comparator which produces the driving signal of the high-speed switch integrated circuit ADG201AKN. This switch allows to change the value of a feedback resistor of an OP-AMP in inverting configuration. In fact, if the switch

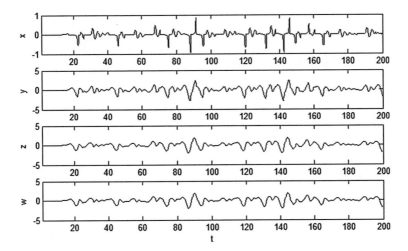

Fig. 3.37 Numerical simulations of Eq. (3.38): trend of the state variables X, Y, Z, and W

is open $R_f = 10$ kΩ, otherwise $R_f = 5$ kΩ. The gain of this block, thus, changes according to the state of the switch. The two feedback resistors are chosen in order to implement the parameters $c = \frac{\frac{R_{31}R_{32}}{R_{31}+R_{32}}}{R_{33}}$ and $d = \frac{R_{32}}{R_{33}}$ of the memristor characteristic. The output of this block corresponds to the output $XH(w)$ of the memristive device.

The whole electrical scheme for the memristive circuit is shown in Fig. 3.40. The circuit obeys to the following equations:

$$\begin{cases} C_1 R_6 \frac{dX}{d\tau} = -X + \frac{R_1}{R_3}X + \frac{R_1}{R_4}Y - \frac{R_1}{R_2}V_M \\ C_2 R_{12} \frac{dY}{d\tau} = -Y + \frac{R_7}{R_8}Z + \frac{R_7}{R_9}Y - \frac{R_7}{R_{10}}X \\ C_3 R_{17} \frac{dZ}{d\tau} = -Z + \frac{R_{13}}{R_{14}}Z - \frac{R_{13}}{R_{15}}Y \\ C_4 R_{22} \frac{dW}{d\tau} = -W + \frac{R_{18}}{R_{21}}W + \frac{R_{18}}{R_{20}}X \end{cases} \qquad (3.39)$$

where $\kappa = \frac{1}{R_6 C_1} = \frac{1}{R_{12}C_2} = \frac{1}{R_{17}C_3} = \frac{1}{R_{22}C_4} = 1000$ is the time scaling factor of the circuit and V_M is the output of the memristive device. Other components are chosen in order to match Eq. (3.38), according to the figure caption.

The behavior of the circuit shows an interesting bifurcation scenario leading to chaos, as demonstrated by the attractor reported in Fig. 3.41.

3.7 A Time-Delay Chaotic Circuit

This section deals with the problem of the design and the implementation of a time-delay chaotic circuit. A simple feedback scheme consisting of a nonlinearity, a first order RC circuit, and a time-delay block is used [9]. The time-delay block is implemented by the series of Bessel filters, which are low-pass filters with a maximally flat magnitude and a maximally linear phase response [10], so that the whole

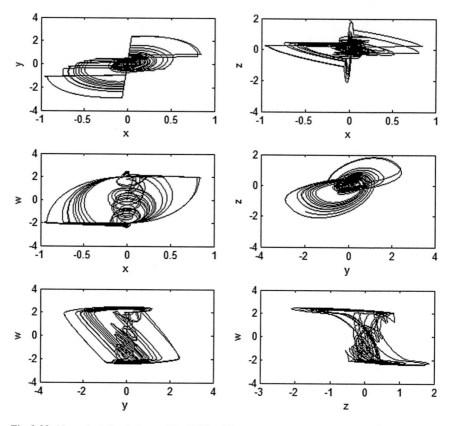

Fig. 3.38 Numerical simulations of Eq. (3.38): different projections of the chaotic attractor, **a** plane X–Y; **b** plane X–Z; **c** plane X–W; **d** plane Y–Z; **e** plane Y–W; **f** plane Z–W

circuit is realized with simple components, like resistors, capacitors, and operational amplifiers. Although a class of circuits based on this topology can be defined, we restrict our analysis to a system described by:

$$\dot{x}(t) = k[-ax(t) - bh(x(t - \tau))] \tag{3.40}$$

with the nonlinearity given by

$$h(x) = \begin{cases} x + 1 & x < 0 \\ 0 & x = 0 \\ x - 1 & x > 0 \end{cases}.$$

In Eq. (3.40) $x(t) \in \mathbb{R}$ is the circuit state variable; $\tau \in \mathbb{R}^+$ is the time-delay; k is a time scaling factor; and a and b are system parameters. For the implemented circuit, we considered the following values of the parameters: $k = 1$, $a = 1$, and $b = 3$.

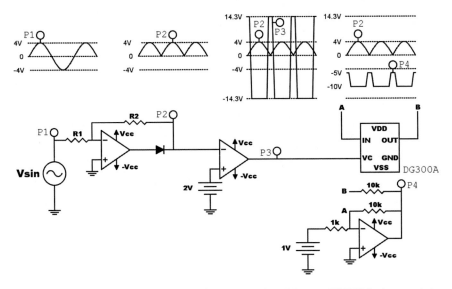

Fig. 3.39 Electronic scheme adopted for the implementation of the term $X H(W)$ in the memristive circuit

The circuit implementing system in Eq. (3.40) is shown in Fig. 3.42. It is governed by the following circuital equation:

$$C_1 R_1 \frac{dX}{d\tau} = -X + \frac{R_5}{R_4} g_1 - \frac{R_5}{R_3} X_\tau \qquad (3.41)$$

where X_τ is the output of the delay chain, while $g_1 = \frac{X_\tau}{|X_\tau|}$ is the output of the nonlinear block, and a time rescaling factor $\kappa = \frac{1}{R_1 C_1} = 1000$ has been introduced.

The experimental observation on the realized circuit (shown in Fig. 3.43) allows also to assess that the delay block, described in Chap. 2, is effective as the chaotic behavior shown in Fig. 3.44 clearly demonstrates.

3.8 The Duffing System

The Duffing system [11] consists of the following nonautonomous set of two differential equations:

$$\begin{cases} \dot{x} = y \\ \dot{y} = x - x^3 - dy + g \sin(\omega t) \end{cases} \qquad (3.42)$$

where the parameters of the system are indicated below:

$$d = 0.25, \, g = 0.3, \, \omega = 1 \qquad (3.43)$$

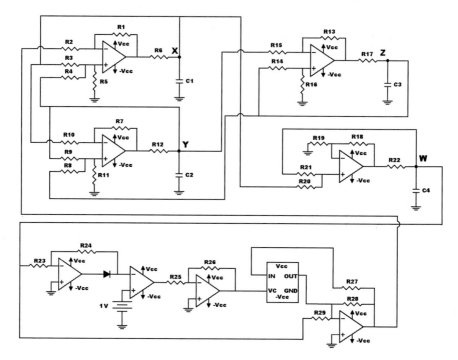

Fig. 3.40 Circuital implementation of the memristive circuit as in Eq. (3.38). Components: $R_1 = 240\,k\Omega$, $R_2 = 60\,k\Omega$, $R_3 = 240\,k\Omega$, $R_4 = 150\,k\Omega$, $R_5 = 100\,k\Omega$, $R_6 = 10\,k\Omega$, $R_7 = 180\,k\Omega$, $R_8 = 720\,k\Omega$, $R_9 = 180\,k\Omega$, $R_{10} = 71.5\,k\Omega$, $R_{11} = 80\,k\Omega$, $R_{12} = 10\,k\Omega$, $R_{13} = 221\,k\Omega$, $R_{14} = 133\,k\Omega$, $R_{15} = 54.9\,k\Omega$, $R_{16} = 66.5\,k\Omega$, $R_{17} = 10\,k\Omega$, $R_{18} = 26\,k\Omega$, $R_{19} = 10\,k\Omega$, $R_{20} = 10\,k\Omega$, $R_{21} = 10\,k\Omega$, $R_{22} = 10\,k\Omega$, $R_{23} = 10\,k\Omega$, $R_{24} = 1\,k\Omega$, $R_{25} = 1\,k\Omega$, $R_{26} = 309\Omega$, $R_{27} = 10\,k\Omega$, $R_{28} = 1\,k\Omega$, $R_{29} = 1\,k\Omega$, $C_1 = C_2 = C_3 = C_4 = 100$ nF, ADG201AKN analog switch, 1N4148 diode, $V_{cc} = 15$ V

The system has been first simulated with *MATLAB*® and the results of the numerical simulation have been shown in Figs. 3.45 and 3.46.

From the analysis of the trajectories of the system, we understand that it is not necessary to scale the system for an electronic implementation. The basic block used in our electronic implementation is the Miller integrator configuration. The whole circuit is shown in Fig. 3.47.

The circuital equations associated to the implementation of the Duffing system are the following:

$$\begin{cases} C_1 R_2 \frac{dX}{d\tau} = y \\ C_2 R_{10} \frac{dY}{d\tau} = -\frac{R_9}{R_5} \frac{x^3}{10} + \frac{R_9}{R_6} x + \frac{R_9}{R_7} V_{\sin} - \frac{R_9}{R_8} y \end{cases} \quad (3.44)$$

where a temporal rescaling $\kappa = \frac{1}{C_1 R_2} = \frac{1}{C_2 R_{10}} = 100000$ has been introduced, $V_{\sin} = \sin(\Omega\tau)$ is the sinusoidal driving signal obtained from a waveform generator,

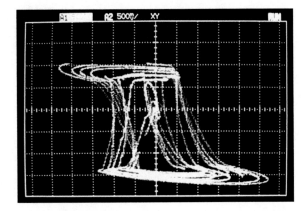

Fig. 3.41 Experimental results: attractor shown by the memristive circuit. Horizontal axis: 500 mV/div; vertical axis: 500 mV/div

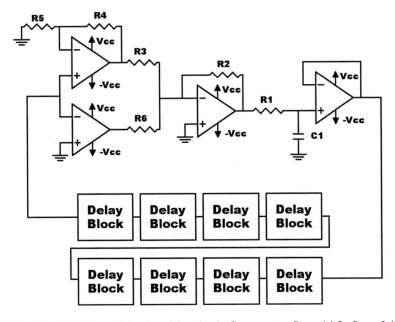

Fig. 3.42 Electrical scheme of the time-delay circuit. Components: $R_1 = 1 \, \text{k}\Omega$, $R_2 = 3.4 \, \text{k}\Omega$, $R_3 = 3.14 \, \text{k}\Omega$, $R_4 = 1.8 \, \text{k}\Omega$, $R_5 = 1 \, \text{k}\Omega$, $R_6 = 7.85 \, \text{k}\Omega$, $C_1 = 1 \, \mu\text{F}$, $V_{cc} = 9 \, \text{V}$

and $\Omega = \kappa\omega$. Matching equation (3.44) with the mathematical model in Eq. (3.42) allows to obtain the values for the components of the circuit. The Duffing circuit contains an external source V_{sin}, whose amplitude represents a bifurcation parameter, which can be suitably changed for setting the waveform generator. In Fig. 3.48, a realization of the circuit is reported, while Fig. 3.49 shows the typical attractor of the Duffing circuit observed on the oscilloscope.

Fig. 3.43 A picture of the
implemented time-delay circuit

Fig. 3.44 Experimental
results: behavior of the circuit for $n = 8$ cascaded
filters. Attractor in the phase
plane $X(T - \tau) - X(T)$ for
the following values of the
parameters: $a = 1, b = 3$

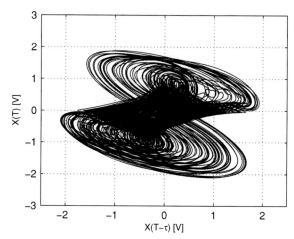

3.9 The Van der Pol Circuit

A further fundamental example of nonlinear systems showing a wide range of nonlinear behaviors is the Van der Pol mathematical model. It has been originally developed
for an electronic oscillator built using vacuum tubes. The Van der Pol oscillator was
originally proposed by the Dutch electrical engineer and physicist Balthasar Van der
Pol while he was working at Philips.

The forced dimensionless equations of Van der Pol circuit are:

$$\begin{cases} \dot{x} = y \\ \dot{y} = a(1 - x^2)y - x + \gamma \sin(\omega t) \end{cases} \tag{3.45}$$

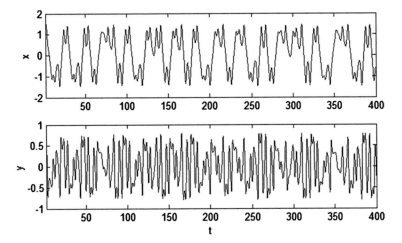

Fig. 3.45 Numerical simulations of Eq. (3.42): trend of the state variables x and y

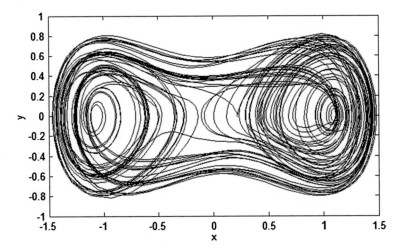

Fig. 3.46 Numerical simulations of Eq. (3.42): chaotic attractor

where the parameters of the system are:

$$a = 8.53 \qquad \gamma = 1.2 \qquad \omega = \frac{2\pi}{10} \qquad (3.46)$$

Before implementing the system of differential equations in a suitable electronic circuit, the oscillations range must be verified simulating Eq. (3.50) with $MATLAB^{\circledR}$. The obtained numerical results are shown in Figs. 3.50 and 3.51.

Even if the oscillations range of the state variables is within the limit imposed by the voltage supply, when the variables are combined in the nonlinear term, values outside the range are obtained. For this reason, a rescaling factor as reported below is applied:

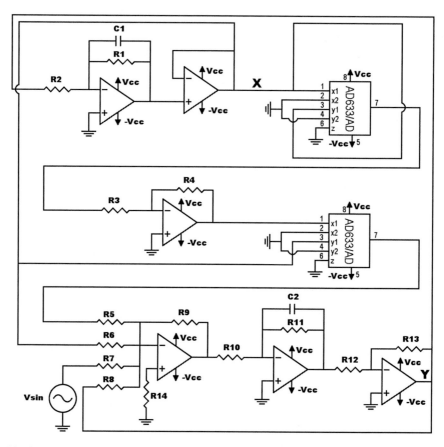

Fig. 3.47 Electronic implementation of Eq. (3.42) Components: $R_1 = 1\,k\Omega$, $R_2 = 100\Omega$, $R_3 = 1\,k\Omega$, $R_4 = 10\,k\Omega$, $R_5 = 100\Omega$, $R_6 = 1\,k\Omega$, $R_7 = 1\,k\Omega$, $R_8 = 4\,k\Omega$, $R_9 = 1\,k\Omega$, $R_{10} = 100\Omega$, $R_{11} = 1\,k\Omega$, $R_{12} = 1\,k\Omega$, $R_{13} = 1\,k\Omega$, $R_{14} = 75\,\Omega$, $C_1 = 100\,nF$, $C_2 = 100\,nF$, $V_{cc} = 9\,V$

Fig. 3.48 A picture of the implemented circuit for Eq. (3.42)

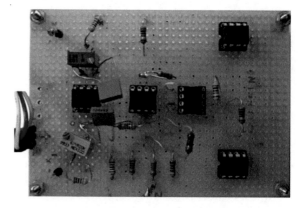

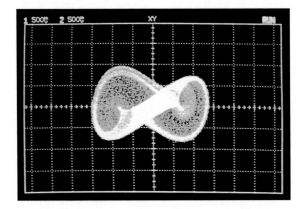

Fig. 3.49 Experimental results: attractor shown by the Duffing circuit. Horizontal axis: 500 mV/div; vertical axis 500 mV/div

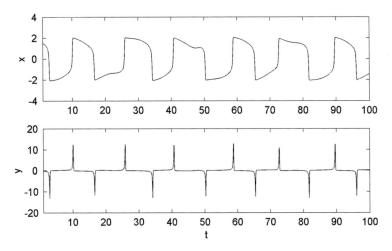

Fig. 3.50 Numerical simulations of Eq. (3.50): trend of the state variables x and y

$$X = k_1 x; \qquad Y = k_2 y \tag{3.47}$$

where

$$k_1 = \frac{1}{3}; \qquad k_2 = \frac{1}{15} \tag{3.48}$$

In this way, a new rescaled equivalent Van der Pol mathematical model is defined by the following equations:

$$\begin{cases} \dot{X} = \frac{15Y}{3} \\ \dot{Y} = a(1 - 9X^2)Y - \frac{X}{5} + \frac{\gamma \sin(\omega t)}{15} \end{cases} \tag{3.49}$$

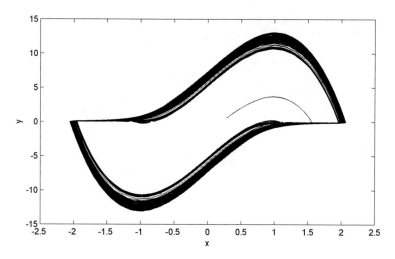

Fig. 3.51   Numerical simulations of Eq. (3.50): chaotic attractor

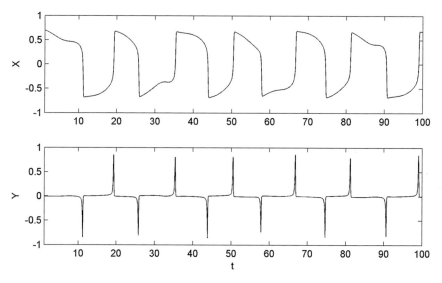

Fig. 3.52   Numerical simulations of Eq. (3.49): trend of the state variables X and Y

The numerical simulation of the rescaled Van der Pol model is shown in Figs. 3.52 and 3.53.

Following the methodology of Chap. 2, an electronic circuit has been designed and implemented using resistors, capacitors, and operational amplifiers. The circuit is shown in Fig. 3.54.

The circuital equations associated to the Van der Pol circuit are the following:

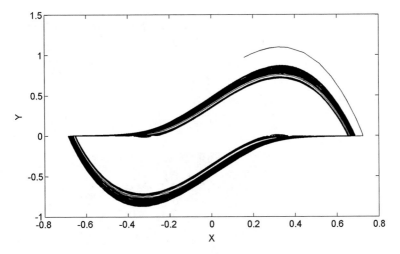

Fig. 3.53 Numerical simulations of Eq. (3.49): chaotic attractor

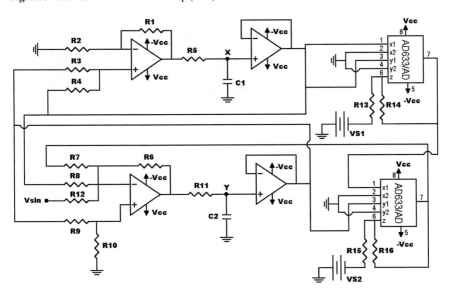

Fig. 3.54 Circuital scheme implementing the Van der Pol model (3.49). Components: $C_1 = 100$ nF, $C_2 = 100$ nF, $R_1 = 100$ kΩ, $R_2 = 20$ kΩ, $R_3 = 20$ kΩ, $R_4 = 100$ kΩ, $R_5 = 10$ kΩ; $R_6 = 100$ kΩ, $R_7 = 1.3$ kΩ, $R_8 = 500$ kΩ, $R_9 = 10.5$ kΩ, $R_{10} = 1.43$ kΩ, $R_{11} = 10$ kΩ, $R_{12} = 100$ kΩ, $R_{13} = 9$ kΩ, $R_{14} = 1$ kΩ, $R_{15} = 9$ kΩ, $R_{16} = 1$ kΩ, $V_{S2} = 44$ mV, $V_{S2} = 44$mV, $V_{cc} = 9$ V, $f_{Vsin} = 100$ Hz, $A_{Vsin} = 0.08 V$

Fig. 3.55 A picture of the
circuit implementing the Van
der Pol model (3.49)

Fig. 3.56 Experimental
results. Attractor of the
Van der Pol circuit. Phase
plane: x–y. Horizontal axis =
200 mV/div, vertical axis =
500 mV/div

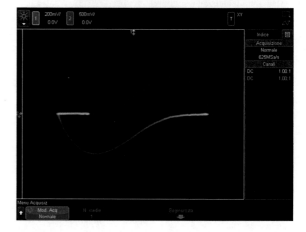

$$\begin{cases} C_1 R_5 \dot{X} = -X - \frac{R_1}{R_3} Y + \frac{R_1}{R_4} X \\ C_2 R_{11} \dot{Y} = -Y - \frac{R_6}{R_7} X^2 Y - \frac{R_6}{R_8} X - \frac{R_6}{R_{12}} V_{\text{sin}} + \frac{R_6}{R_9} Y \end{cases} \qquad (3.50)$$

In Fig. 3.55, a picture of the realized circuit is shown. Experimental results are
reported in Fig. 3.56.

3.10 The Dissipative Nonautonomous Chaotic Circuit

In this section, the implementation of a dissipative oscillator with a nonlinearity
realized by PWL function is described. In the proposed circuit [12], the dissipative
terms introduced by the Miller integrators are exploited to realize specific terms of the
mathematical model. The dimensionless equations describing the system dynamics
are the following:

$$\begin{aligned} \dot{x} &= y - a_1 x \\ \dot{y} &= -bx - a_2 y + \sin(\omega t) + s(x) \end{aligned} \qquad (3.51)$$

where $s(x) = \frac{1}{2}(|5x + 1| - |5x - 1|)$.

The numerical simulation performed integrating Eq. (3.51) with $a_1 = a_2 = 0.01$,
$b = 1$, $\omega = 0.2$ rad/s is reported in Figs. 3.57 and 3.58. The observation of the

Fig. 3.57 Numerical
simulations of Eq. (3.51):
trend of the state variables
x and y

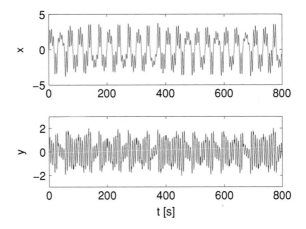

Fig. 3.58 Numerical
simulations of Eq. (3.51):
chaotic attractor

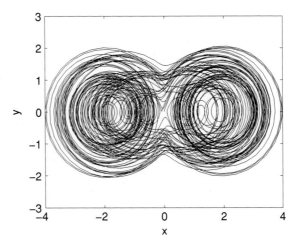

oscillation range for both state variables allows to assess that they remain within
the limits imposed by the voltage supply of ± 9 V. Even if an amplitude rescaling is
not needed, in the view of a circuital implementation, it should be noticed that the
actual behavior of the system can be characterized only with long observation. This
can be solved applying a temporal rescaling in the circuit dynamics, following the
guidelines described in Chap. 2.

Furthermore, in order to implement the dissipative terms, we notice that they
can be associated with the nonideal behavior of the Miller integrators. In fact, the
transfer functions of such blocks do not have a zero pole (as in the integrator ideal
case). More precisely, the low frequency pole of such transfer functions is placed in
$s = -\frac{1}{R_7 C_1} = \frac{1}{R_{11} C_2}$ which implies a dissipative term. Hence, the equations of the
circuit shown in Fig. 3.59 read as follows:

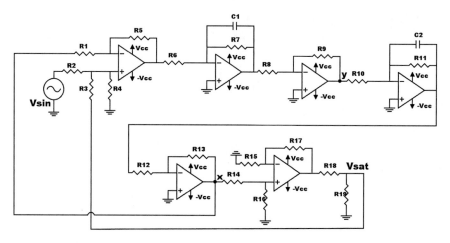

Fig. 3.59 Circuital implementation of Eq. (3.51). Components: $R_1 = 1\,\text{k}\Omega$, $R_2 = 2\,\text{k}\Omega$, $R_3 = 2\,\text{k}\Omega$, $R_4 = 1\,\text{k}\Omega$, $R_5 = 1\,\text{k}\Omega$, $R_6 = 3.3\,\text{k}\Omega$, $R_7 = 330\,\text{k}\Omega$, $R_8 = 1\,\text{k}\Omega$, $R_9 = 1\,\text{k}\Omega$, $R_{10} = 3.3\,\text{k}\Omega$, $R_{11} = 330\,\text{k}\Omega$, $R_{12} = 1\,\text{k}\Omega$, $R_{13} = 1\,\text{k}\Omega$, $R_{14} = 5\,\text{k}\Omega$, $R_{15} = 5\,\text{k}\Omega$, $R_{16} = 470\,\text{k}\Omega$, $R_{17} = 470\,\text{k}\Omega$, $R_{18} = 16.5\,\text{k}\Omega$, $R_{19} = 2\,\text{k}\Omega$, $C_1 = 100\,\text{nF}$, $C_2 = 100\,\text{nF}$, $V_{cc} = 9\,\text{V}$

Fig. 3.60 Picture of the implemented circuit realizing system (3.51)

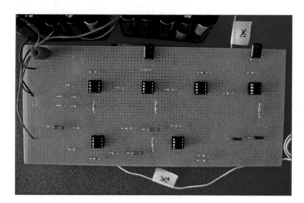

$$C_1 R_6 \frac{dX}{d\tau} = \frac{R_{11}R_{13}}{R_{10}R_{12}} Y - \frac{1}{\kappa R_{11}C_2} X$$
$$C_2 R_{10} \frac{dY}{d\tau} = \frac{R_7 R_9}{R_6 R_8}\left(-\frac{R_5 R_{13}}{R_1 R_{12}} X + \frac{R_5}{R_2}\sin(\Omega\tau) + \frac{R_5}{R_3} V_{PWL} - \frac{1}{\kappa R_7 C_1} Y\right) \tag{3.52}$$

where $\kappa = \frac{1}{C_1 R_6} = \frac{1}{C_2 R_{10}} = 3000$ is the adopted temporal rescaling, $\Omega = \kappa\omega$ and V_{PWL} is the PWL nonlinearity implemented exploiting OP-AMP saturations. The values of the components are reported in the caption of Fig. 3.59 and match Eq. (3.52) with Eq. (3.51). In particular, we have set $\frac{R_9}{R_8} = 1$, $\frac{R_{13}}{R_{12}} = 1$, $\frac{R_5 R_{13}}{R_1 R_{12}} = 1$, $\frac{R_5}{R_2} = \frac{1}{2}$, and $\frac{R_5}{R_3} = \frac{1}{2}$. The dissipative terms are $\frac{1}{\kappa R_{11}C_2} = \frac{1}{\kappa R_7 C_1} = 0.01$.

The behavior of the implemented circuit, reported in Fig. 3.60, confirms the rich variety of dynamics ranging from limit cycles to chaos, which can be observed

Fig. 3.61 Experimental results of the circuit implementing Eq. (3.51): attractor in the phase plane x–y. Horizontal axis = 500 mV/div, vertical axis = 500 mV/div

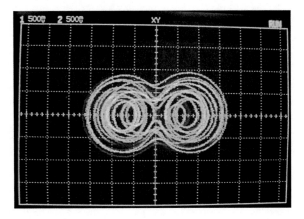

varying the frequency of the driving sinusoidal signal V_{sin} supplied by an external function generator. As an example, the chaotic attractor obtained when $f = \frac{\Omega}{2\pi} =$ 95 Hz can be observed in the picture taken from the oscilloscope reported in Fig. 3.61.

References

1. Sprott JC (2011) A new chaotic jerk circuit. IEEE Trans Circuits Syst II 58(4):240–243
2. Arena P, Baglio S, Fortuna L, Manganaro G (1995) Chua's circuit can be generated by CNN cells. IEEE Trans Circuits Syst I 42(2):123–125
3. Fortuna L, Frasca M, Xibilia MG (2009) Chua's circuit implementations: yesterday, today, tomorrow. World Scientific, Singapore
4. Lorenz EN (1963) Deterministic nonperiodic flow. J Atmos Sci 20(2):130–141
5. Hindmarsh JL, Rose RM (1984) A model of neuronal bursting using three coupled first order differential equations. Proc R Soc London Ser B 221:87–102
6. Mullin TJ (1993) Nonlinear phenomena and chaos in fluid flows. In: Dendy R (ed) Plasma physics: an introductory course. Cambridge University Press, Cambridge
7. Chua LO (1971) Memristor—the missing circuit element. IEEE Trans Circuit Theory 18:507–519
8. Buscarino A, Fortuna L, Frasca M, Gambuzza LV, Sciuto G (2012) Memristive chaotic circuits based on cellular nonlinear networks. Int J Bifurcat Chaos 22(3):1250070-1-13
9. Buscarino A, Fortuna L, Frasca M, Sciuto G (2011) Design of time-delay chaotic electronic circuits. IEEE Trans Circuits Syst I 58(8):1888–1896
10. Daryanani G (1976) Principles of active network synthesis and design. Wiley, Singapore
11. Sprott JC (2003) Chaos and time-series analysis. Oxford University Press, Oxford
12. Buscarino A, Fortuna L, Frasca M (2009) A new CNN-based chaotic circuit: experimental results. Int J Bifurcat Chaos 19(8):2609–2617

Chapter 4
FPAA-Based Implementation of Chaotic Circuits

Abstract In this chapter, three examples of the use of FPAA for the implementation of chaotic circuits are given. In the three examples, the design is based on the procedure of Chap. 2, by taking now into account the bounds of the internal voltage signals of the FPAA board. The implementation is based on a FPAA device produced by ANADIGM: the AN221E04 FPAA [1].

Keywords Chaos · Chaotic circuits · FPAA

4.1 The FPAA-Based Chua's Circuit

In the design of a FPAA-based Chua's circuit, the first step is to rescale the system, so that, the internal voltage signals are within the bounds of the FPAA. For the AN221E04 FPAA these are ± 2 V. These values drive the choice of the scaling factor. In particular, the scaled variables are

$$\begin{cases} X = \dfrac{x}{k} \\ Y = y \\ Z = \dfrac{z}{k} \end{cases} \qquad (4.1)$$

with $k = 2$. The equations of the Chua's circuit, therefore, take the following form:

A. Buscarino et al., *A Concise Guide to Chaotic Electronic Circuits*,
SpringerBriefs in Applied Sciences and Technology,
DOI: 10.1007/978-3-319-05900-6_4, © The Author(s) 2014

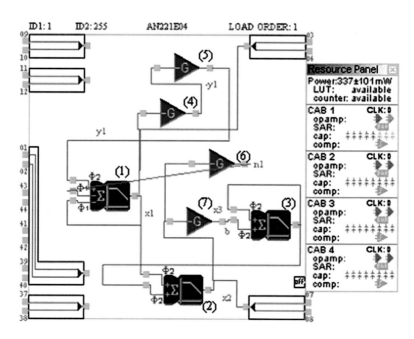

Fig. 4.1 Scheme of the Chua's circuit implemented by an FPAA

$$
\begin{cases}
\dot{X} = \dfrac{\alpha}{k}[Y - h(kX)] \\
\dot{Y} = kX - Y + kZ \\
\dot{Z} = -\dfrac{\beta}{k}Y.
\end{cases}
\tag{4.2}
$$

Figure 4.1 shows the blocks, used inside the environment tool provided by the ANADIGM company, for implementing the Chua's circuit [2]. Each block is characterized by a number, from 1 to 7, and the function of each single block is described below. The type of the block refers to the terminology adopted in the ANADIGM development tool.

Blocks (1)–(3) are of the type SUMFILTER, while blocks (4)–(7) are GAININV blocks. The output of each of the SUMFILTER blocks is one of the three state space variables X, Y, and Z; these blocks perform algebraic sum of the input signals and the integration for each state variable. Some of the circuit gains were implemented directly inside the SUMFILTER blocks, while others were fixed using the GAININV

Table 4.1 Parameters used in FPAA-based implementation of the Chua's circuit

CAM type	CAM number	Tuned parameter	Parameter value
	1	Corner frequency	0.4
		Gain 1	2
		Gain 2	1
		Gain 3	1.5
SUMFILTER	2	Corner frequency	0.4
		Gain 1	2
		Gain 2	2
	3	Corner frequency	0.4
		Gain 1	1
		Gain 2	1
	4	Gain	4
GAININV	5	Gain	0.5
	6	Gain	g
	7	Gain	6

blocks. This was due to the limited range of gains settings in the SUMFILTER block (too small for implementing the parameters α and β) and to the accuracy of the values implemented which is higher if GAININV blocks are used.

The GAININV blocks (4) and (5) are used to implement the output nonlinearity, by exploiting the 2V saturation of the ANADIGM blocks. The values of the circuit parameters were chosen to implement the well-known double-scroll chaotic attractor. The used parameters are reported in Table 4.1. For the GAININV block (6), a gain g equal to $g = 3.18$ has been used. This leads to the experimental results reported in Figs. 4.2 and 4.3. Figure 4.2 shows the projection of the double-scroll chaotic attractor in the $X - Y$ plane, while Fig. 4.3 shows the state variables X and Y. By changing the value of g, the period-doubling route to chaos can be observed. This can be also done in an interactive way. It is, in fact, possible to create a graphic user interface with a slide controlling the parameter g of the Chua's circuit, so that a low-cost introductory kit on the basic properties of chaotic circuits for educational purposes may be easily built.

4.2 The FPAA Multiscroll Circuit

The double-scroll chaotic attractor of the Chua's circuit is not the only attractor characterized by scrolls, that is, orbits around one of the unstable equilibrium points of the circuit. Attractors with an arbitrary number of scrolls, called *multiscroll attractors*, have been also obtained with several approaches [3, 4]. Stair circuit, hysteresis circuit, and saturated circuit are the three kinds of basic circuits used for creating multiscroll chaotic attractors. In this example, we take into account the approach

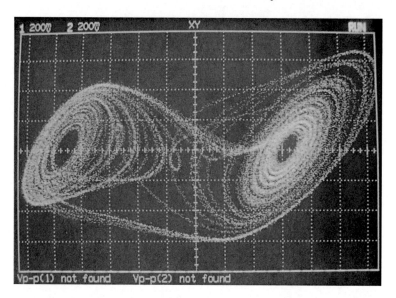

Fig. 4.2 Experimental results of the FPAA-based implementation of the Chua's circuit: projection of the double-scroll attractor in the $X-Y$ plane ($g = 3.18$)

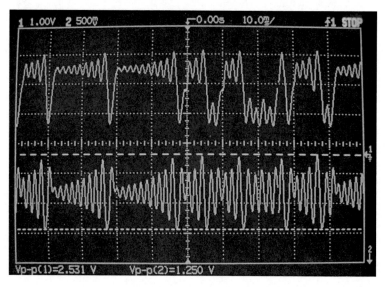

Fig. 4.3 Experimental results of the FPAA-based implementation of the Chua's circuit: trend of the state variables X (*up*) and Y (*bottom*) ($g = 3.18$)

based on saturated function series discussed in [4]. According to this approach, the multiscroll system is described by:

$$\begin{cases} \dot{x} = y - \dfrac{d_2}{b} f(y; k_2, h_2, p_2, q_2) \\ \dot{y} = z \\ \dot{z} = -ax - by - cz + d_1 f(x; k_1, h_1, p_1, q_1) + d_2 f(y; k_2, h_2, p_2, q_2) \end{cases} \tag{4.3}$$

where

$$f(x; k, h, p, q) = \begin{cases} k(2q + 1) & \text{if } x > qh + 1 \\ k(x - ih) + 2ik & \text{if } |x - ih| \leq 1, \ -p \leq i \leq q \\ k(2i + 1) & \text{if } ih + 1 < x < (i + 1)h - 1, \ -p \leq i \leq q - 1 \\ -k(2p + 1) & \text{if } x < -ph - 1 \end{cases} \tag{4.4}$$

where the parameters k, h, p, and q represent the saturated slope, the distance between two consecutive saturated slopes, and two integer constants controlling the number of scrolls in negative and positive directions of the variable. We focus on a 3×3 grid scroll chaotic attractor by selecting the following values for the parameters: $a = b = c = 0.7, d_1 = d_2 = 0.7, k_1 = k_2 = 10, h_1 = h_2 = 20, p_1 = p_2 = 0,$ $q_1 = q_2 = 1$. Since the dynamic range of the system variables (the variables x and y have peak-to-peak oscillations equal to 40 units, while the variable z variable to 10 units) is larger than that allowed by the FPAA device, the following rescaling was adopted:

$$X = \frac{x}{K_1}; \qquad Y = \frac{y}{K_2}; \qquad Z = \frac{z}{K_3} \tag{4.5}$$

with $K_1 = K_2 = 10$ and $K_3 = 5$, thus transforming Eq. (4.3) into

$$\begin{cases} \dot{X} = \dfrac{K_2}{K_1} Y - \dfrac{d_2}{bK_1} f(K_2 Y; k_2, h_2, p_2, q_2) \\ \dot{Y} = \dfrac{K_3}{K_2} Z \\ \dot{Z} = -a\dfrac{K_1}{K_3} x - b\dfrac{K_2}{K_3} y - cz + \dfrac{d_1}{K_3} f(K_1 X; k_1, h_1, p_1, q_1) \\ \qquad + \dfrac{d_2}{K_3} f(K_2 Y; k_2, h_2, p_2, q_2). \end{cases} \tag{4.6}$$

The electrical scheme used for implementing Eq. (4.6) is shown in Fig. 4.4. Two FPAA boards, connected together, are required, since the number of programmable blocks in each is not enough to reproduce Eq. (4.6). The design follows the guidelines described in Chap. 2: the only thing which is worth to mention is the implementation of the nonlinearity, which is done by two user-defined blocks called in the development program as transfer functions, and indicated as TF4 (devoted to implement $f(Y; k_2, h_2, p_2, q_2)$) and TF5 (realizing $f(X; k_1, h_1, p_1, q_1)$). These blocks implement a user-defined voltage transfer function with 256 quantization steps, as

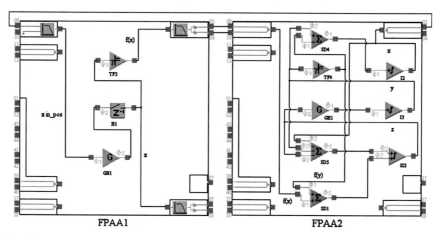

Fig. 4.4 Schematic of the FPAA implementation of Eq. (4.6)

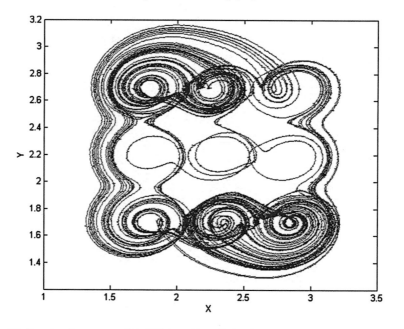

Fig. 4.5 The chaotic attractor of the FPAA multiscroll circuit

specified through a Lookup Table edited by the user. The chaotic attractor, experimentally obtained, is shown in Fig. 4.5, which shows 160000 recorded samples with a sampling frequency of 100 kHz (which correspond to a time duration of 1.6 s).

4.3 A Circuit Implementing CO_2 Laser Dynamics

As a third example of FPAA-based circuits, we discuss a model of a CO_2 laser [5]. In CO_2 lasers, there is not a single mode field interacting with a resonant molecular transition, but more complicated population transfers occur that need a four-level system to be modeled. The model consists of the following six equations:

$$\dot{x}_1 = k_0 x_1 (x_2 - 1 - k_1 \sin^2 x_6)$$
$$\dot{x}_2 = -\Gamma_1 x_2 - 2k_0 x_1 x_2 + \gamma x_3 + x_4 + P_0$$
$$\dot{x}_3 = -\Gamma_1 x_3 + \gamma x_2 + x_5 + P_0$$
$$\dot{x}_4 = -\Gamma_2 x_4 + \gamma x_5 + z x_2 + z P_0 \qquad (4.7)$$
$$\dot{x}_5 = -\Gamma_2 x_5 + \gamma x_4 + z x_3 + z P_0$$
$$\dot{x}_6 = -\beta x_6 + \beta B_0 - \beta \frac{R x_1}{1 + \alpha x_1}$$

where x_1 represents the photon number proportional to the laser intensity, x_2 is proportional to the laser inversion, x_3 is proportional to the sum of the populations of the laser resonant levels, x_4 and x_5 are, respectively, proportional to the difference and sum of the populations of the rotational manifolds coupled to the resonant levels, and x_6 is a term proportional to the feedback voltage acting on the cavity loss. The system parameters are fixed to: $k_0 = 28.5714$, $k_1 = 4.5556$, $\gamma = 0.05$, $P_0 = 0.016$, $\alpha = 32.8767$, $\beta = 0.4286$, $\Gamma_1 = 10.0643$, $\Gamma_2 = 1.0643$. For these values, the laser intensity is characterized by spikes and chaos is manifest in the form of irregular spike intervals, associated to the perturbation of the homoclinic orbit of the system.

Equations (4.7) are first rewritten in a suitable form for FPAA-based implementation. The nonlinearity cannot be directly implemented by standard FPAA blocks and it has been simplified by taking into account that in the dynamic range of the variables $\sin^2 x \simeq x^2$. The equations are then rescaled using the variables:

$$X_1 = \frac{x_1}{k_1}; \quad X_2 = \frac{x_2}{k_2}; \quad X_3 = \frac{x_3}{k_3}; \quad X_4 = \frac{x_4}{k_4}; \quad X_5 = \frac{x_5}{k_5}; \quad X_6 = \frac{x_6}{k_6}$$
$$(4.8)$$

with $k_1 = 1/200$, $k_2 = 1$, $k_3 = 1$, $k_4 = 10$, $k_5 = 10$, $k_6 = 1$, so that the rescaled system is:

$$\dot{X}_1 = k_0 X_1 (X_2 - 1 - k_1 X_6^2)$$
$$\dot{X}_2 = -\Gamma_1 X_2 - 2k_0 k_1 X_1 X_2 + \gamma X_3 + k_4 X_4 + P_0$$
$$\dot{X}_3 = -\Gamma_1 X_3 + \gamma X_2 + k_5 X_5 + P_0$$
$$\dot{X}_4 = -\Gamma_2 X_4 + \gamma \frac{k_5 X_5}{k_4} + z X_2 / k_4 + z P_0 / k_4 \qquad (4.9)$$

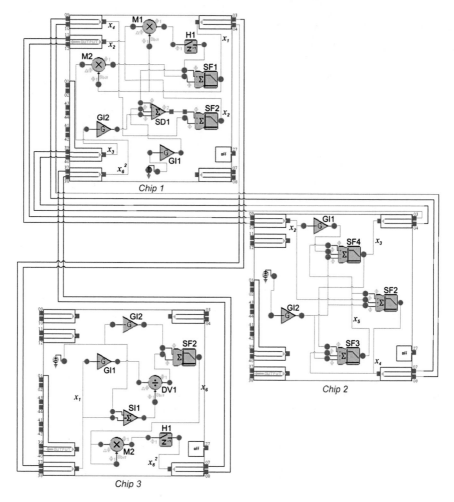

Fig. 4.6 Scheme of the circuit implementing Eq. (4.9)

$$\dot{X}_5 = -\Gamma_2 X_5 + \gamma \frac{k_4 X_4}{k_5} + z X_3/k_5 + z P_0/k_5$$

$$\dot{X}_6 = -\beta X_6 + \beta B_0 - \beta \frac{R X_1 k_1}{1 + \alpha k_1 X_1}.$$

The scheme of the circuit is reported in Fig. 4.6, where three FPAA boards have been connected together in order to reach the number of programmable blocks needed for the implementation. Experimental results are reported in Figs. 4.7 and 4.8, where all the data have been acquired by using a data acquisition board (National Instruments AT-MIO 1620E) with a sampling frequency $f_s = 50\,\text{kHz}$ and subsequently plotted by using MATLAB®. In particular, Fig. 4.7 shows the trend

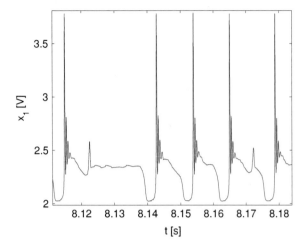

Fig. 4.7 Experimental results of the circuit implementing Eq. (4.9): trend of the variable X_1

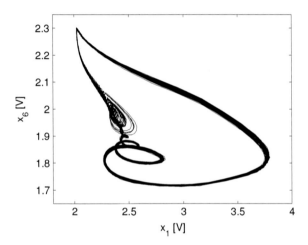

Fig. 4.8 Experimental results of the circuit implementing Eq. (4.9): projection of the attractor on the plane X_1–X_6

of the state variable X_1, while Fig. 4.8 the projection of the attractor on the plane X_1–X_6.

References

1. http://www.anadigm.com
2. Caponetto R, Di Mauro A, Fortuna L, Frasca M (2005) Field programmable analog arrays to implement programmable Chua's circuit. Int J Bifurcat Chaos 15(5):1829–1836

3. Suykens JAK, Vandewalle J (1993) Generation of n-double scrolls ($n = 1; 2; 3; 4; \ldots$). IEEE Trans Circuits Syst I 40:861–867
4. Lu J, Chen G, Yu X, Leung H (2004) Design and analysis of multiscroll chaotic attractors from saturated function series. IEEE Trans Circuits Syst I 51(12):2476–2490
5. Pisarchik AN, Meucci R, Arecchi FT (2001) Theoretical and experimental study of discrete behavior of Shilnikov chaos in a CO_2 laser. Eur Phys J D 13:385–391

Chapter 5
Synchronization of Chaotic Circuits

Abstract Aim of this chapter is to show techniques and examples of synchronization of chaotic circuits. Two cases are dealt with: synchronization of nominally identical chaotic circuits and synchronization of circuits with parametric or structural differences. In the first case, the circuits are assumed to be regulated by the same dynamics but having slightly different parameters (in the limit of component tolerances), while in the second case circuits with different dynamical behaviors either due to a different parameter or to different dynamical equations are considered.

Keywords Chaos · Chaotic circuits · Synchronization

5.1 Synchronization of Identical Chaotic Systems

Synchronization is the process through which two or more coupled circuits adjust a given property towards a common feature, thanks to a form of coupling or common external forcing [1, 2]. In nonlinear dynamical systems, and in particular in chaotic systems, the case in which the state variables of two (or more) systems follows the same trajectory is referred to as complete synchronization. In this chapter, complete synchronization of two diffusively coupled circuits is discussed from the analytical and experimental points of view through an illustrative case study: synchronization of two Chua's circuits. The conditions under which complete synchronization is ensured are derived by applying a strategy based on the Master Stability Function. Diffusive coupling is then implemented by connecting the two circuits with a passive resistor.

Several approaches allow to design the coupling between the circuits so that a stable synchronous behavior is obtained: negative feedback [3], sporadic driving [4], active-passive decomposition [5, 6], diffusive coupling and some other hybrid methods [7]. Synchronization can be achieved with either unidirectional or bidirectional coupling. In the case of two coupled dynamical units, if the coupling is unidirectional, one chaotic system remains unaltered while forcing the other to follow its dynamics; if the coupling is bidirectional, both systems are influenced each other.

A. Buscarino et al., *A Concise Guide to Chaotic Electronic Circuits*,
SpringerBriefs in Applied Sciences and Technology,
DOI: 10.1007/978-3-319-05900-6_5, © The Author(s) 2014

This latter case can be implemented in an electronic circuit simply through a resistor connecting two corresponding state variables of the chaotic circuits. The use of diffusive coupling in Chua's circuits has been considered in [5] where it is shown that a diffusive coupling can produce the instability of fixed points, causing the emergence of chaos. Moreover, the synchronization of two Chua's circuits coupled in a bidirectional diffusive scheme has been considered in [8] where a sufficient condition for the onset of synchronization, based on the evaluation of the stability of the linearized error system, is given.

In this chapter, a strategy based on the Master Stability Function (MSF) is used to derive the analytical conditions for the occurrence of synchronization in several possible cases (i.e. scalar coupling with each of the state variables). Experiments have been performed on the implementation of two Chua's circuits obtained following the procedure described in Chap. 2. From the observation of real circuits it is possible to state that, even if in the real case the two circuits are not exactly identical, diffusive coupling does work properly and synchronization occurs. Synchronization of two Chua's circuits is thus obtained with a simple element like a resistor, that is, without any active component in the coupling.

5.1.1 Master Stability Function Based Strategy

The Master Stability Function (MSF), introduced in [9], is an efficient tool to evaluate the conditions under which N identical oscillators can be synchronized when coupled through a network configuration admitting an invariant synchronization manifold. According to the MSF formalism, the dynamics of each node is modeled as

$$\dot{x}^i = F(x^i) - \sigma \sum_{j=1}^{N} G_{ij} H(x^j) \tag{5.1}$$

where $x^i \in \mathbb{R}^n, i = 1 \cdots N, \dot{x}^i = F(x^i)$ represents the uncoupled dynamics of each node, σ is the coupling coefficient, $H : \mathbb{R}^n \to \mathbb{R}^n$ the coupling function and $G = [G_{ij}]$ is a zero-row sum matrix modeling the coupling network. Linearizing Eq. (5.1) around the synchronous state and performing a diagonalization of the resulting time-varying linear equation, a generic variational equation is obtained:

$$\dot{\zeta} = [DF - (\alpha + i\beta)DH]\zeta \tag{5.2}$$

where DF and DH represent the Jacobian of $F(x^i)$ and $H(x^j)$ computed around the synchronous state. Calculating the maximum conditional Lyapunov exponent λ_{max} of Eq. (5.2) as a function of α and β, the necessary conditions for synchronization can be derived. The function $\lambda_{max} = \lambda_{max}(\alpha + i\beta)$ is independent of the specific network topology and it is called the MSF. Hence, the stability of the synchronization manifold in a given network is evaluated by computing the eigenvalues γ_h (with $h = 2 \cdots N$)

of the matrix G and checking if the sign of λ_{max} at the points $\alpha + i\beta = \sigma\gamma_h$ is either negative (corresponding to a stable eigenmode) or positive (corresponding to an unstable eigenmode). If all associated eigenmodes with $h = 2 \cdots N$ are stable, then the synchronous state is stable at the given coupling strength σ.

The approach based on the MSF can be applied to any network admitting a synchronization manifold. We restrict it now to the case of two diffusively coupled circuits, for which experimental results will be also included. In fact, when diffusive coupling between two nodes is considered, the coupling matrix becomes

$$G = \begin{bmatrix} 1 & -1 \\ -1 & 1 \end{bmatrix}$$

that is, a matrix with all real eigenvalues, i.e. $\gamma_1 = 0$ and $\gamma_2 = 2$. In this case, the MSF can be computed as function of α only (this condition holds whenever the coupling matrix is symmetric, that is, for instance when the network is undirected). The functional dependence of λ_{max} on α generally gives rise to three different cases [10]. The first case (type I MSF) is the case in which λ_{max} is positive $\forall\alpha$ and, thus, a stable synchronous state cannot be obtained. In the second case (type II) above a threshold value, say α_c, λ_{max} assumes negative values. Therefore, in this case, a high enough coupling strength always leads to a stable synchronous state: a stable synchronous state is obtained if the coupling strength satisfies $\alpha > \alpha_c$. In the third case (type III), λ_{max} is negative only in some interval of α: a stable synchronous state is obtained if the coupling strength lies in such interval.

5.1.2 Synchronization of Two Diffusively Coupled Chua's Circuits

In this section, as a case study of two identical circuits subjected to differences generated by component tolerances, synchronization of two diffusively coupled Chua's circuits is considered [11]. The aim of the experiments reported here is to use the MSF based approach to assess analytically the conditions for the stability of the synchronous state, and consequently derive the diffusive gain, thus enabling to ensure the onset of synchronization.

We start by reporting the dimensionless equations for two diffusively coupled Chua's circuits:

$$\begin{cases} \dot{x}_1 = \bar{\alpha}(y_1 - h(x_1)) + \sigma\delta_x(x_2 - x_1) \\ \dot{y}_1 = x_1 - y_1 + z_1 + \sigma\delta_y(y_2 - y_1) \\ \dot{z}_1 = -\beta y_1 + \sigma\delta_z(z_2 - z_1) \\ \dot{x}_2 = \bar{\alpha}(y_2 - h(x_2)) + \sigma\delta_x(x_1 - x_2) \\ \dot{y}_2 = x_2 - y_2 + z_2 + \sigma\delta_y(y_1 - y_2) \\ \dot{z}_2 = -\beta y_2 + \sigma\delta_z(z_1 - z_2) \end{cases} \tag{5.3}$$

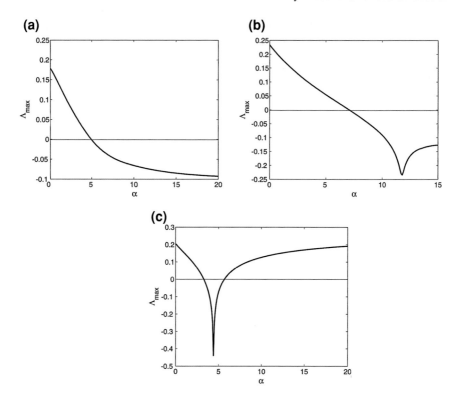

Fig. 5.1 Master Stability Function for the case of diffusively coupled Chua's circuits. Diffusion operates along (**a**) x_1 and x_2, (**b**) y_1 and y_2, (**c**) z_1 and z_2

where the parameter $\bar{\alpha}$ of the Chua's circuit has been relabeled to avoid confusion, and δ_x, δ_y and δ_z can be either one, if the two corresponding state variables are diffusively coupled, or zero otherwise. σ is the coupling coefficient, as previously indicated. In Fig. 5.1, the three MSFs corresponding to the cases in which only a scalar signal is used in the synchronization scheme are reported. In Fig. 5.1a, b, corresponding to the cases $\delta_y = \delta_z = 0$ and $\delta_x = \delta_z = 0$ respectively, it can be noticed that the computed MSF is type II, hence for values above a threshold, the MSF is negative and then the synchronous state is stable. In Fig. 5.1, corresponding to the case $\delta_x = \delta_y = 0$, the MSF is type III, hence synchronization can be achieved only if the coupling strength is chosen in the appropriate interval.

We consider the Chua's circuit implementation of Fig. 3.10 and implement diffusive coupling by introducing a single resistor R_c between two corresponding state variables, as reported in Fig. 5.2. It is easy to show that in this case the coupling strength is given by: $\sigma = \bar{\alpha} R_6 / R_c$. The analysis performed through the MSF allows, in the first two cases, to define for the occurrence of synchronization a lower bound for the coupling strength, and thus an upper bound for the coupling resistor R_c. In particular, from Fig. 5.1a, a value of $\sigma_{cx y2} \approx 5$ can be derived for

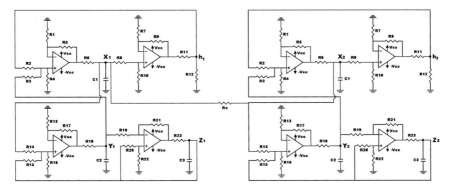

Fig. 5.2 Scheme of the diffusive coupling performed by the variable resistor R_c. The case in which diffusion is implemented between state variables X_1 and X_2 is reported

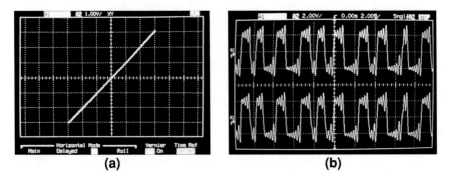

Fig. 5.3 Experimental results: synchronization plot Y_1 versus Y_2 (**a**) and waveforms related to the state variables Z_1 and Z_2 (**b**) for the case in which $\delta_y = \delta_z = 0$

the case of diffusive coupling on the variable x. $\sigma_{cx}\gamma_2 = \gamma_2\bar{\alpha}R_6/R_c > 5$, hence $R_c < \gamma_2\bar{\alpha}R_6/5 \approx 1.25\,k\Omega$.

When considering the diffusive coupling through the second state variable, in the scheme of Fig. 5.2 the coupling resistor R_c is moved between the capacitors implementing Y_1 and Y_2. From Fig. 5.1b, the bound $R_c < \gamma_2 R_{12}/7 \approx 285\,\Omega$ is instead obtained. Although the theory of MSF is developed for identical circuits, when, as in experiments, differences due to component tolerances occur, the approach still holds provided that the differences are small enough.

We now report some illustrative examples of synchronous motion experimentally obtained on the oscilloscope. Figure 5.3 refers to the case of diffusion implemented between the state variables X_1 and X_2, i.e. $\delta_y = \delta_z = 0$: a resistor $R_c = 330\,\Omega$ corresponding to a diffusive constant $\sigma\gamma_2 = \gamma_2\alpha R_6/R_c = 18.98$ at which the MSF has a negative value has been used in this case. In Fig. 5.3a, b two pictures taken from the oscilloscope are reported showing the synchronized behavior of the two diffusively coupled Chua's circuits.

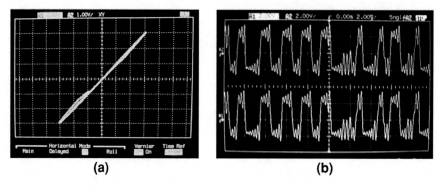

Fig. 5.4 Experimental results: synchronization plot X_1 versus X_2 (**a**) and waveforms related to the state variables Z_1 and Z_2 (**b**) for the case in which $\delta_x = \delta_z = 0$

Figure 5.4 refers to diffusive coupling between the state variables Y_1 and Y_2, i.e. $\delta_y = \delta_z = 0$. Synchronization is obtained with a coupling resistor $R_c = 180\,\Omega$ implementing a diffusive constant $\sigma = \gamma_2 R_{12}/R_c = 11.10$ which leads to a negative value of the MSF, as confirmed by the experimental results reported in Fig. 5.4a, b.

5.2 Synchronization of Non-identical Chaotic Circuits

In this section, synchronization of chaotic circuits with different dynamical behavior is taken into account. A procedure, based on the design of an observer, to assess the necessary conditions for the synchronization of non-identical circuits is discussed. The results obtained are then evaluated through experiments showing the synchronization of chaotic circuits differing either in parameter values or in their dynamical equations. Aim of the experiments is to show the effective synchronization of pairs of non-identical chaotic circuits, i.e. circuits affected by either structural or parametric differences.

Synchronization of chaotic systems affected by uncertainties has been addressed in a limited number of papers. In particular, in [12, 13] the case of parametric mismatches between systems is considered, while in [14–16] the case of structurally different coupled circuits is investigated. More in details, in [12] the synchronization of two Lorenz systems in two different chaotic regions coupled through a master-slave negative feedback scheme [3] is achieved by using three different scalar signals to force the slave dynamics. The phase synchronized motion of a lattice of non-identical Rossler oscillators is considered in [13]. Moreover, dealing with synchronization between non-identical dynamical systems, in [14] an active control strategy is introduced and synchronization of a Chen and a Liu system driven by a Lorenz oscillator is shown. The design procedure of a scalar controller for the synchronization of two non-identical systems is provided in [15] showing the synchronization between a Chua's circuit and a Rossler oscillator. Finally, in [16] a sliding mode controller based on Lyapunov stability theory is introduced. Furthermore, both cases of structural and parametric differences have been considered in

[17] where the analytical conditions to achieve synchronization are given considering a master-slave coupling configuration and sending three different scalar signals.

Here, the strategy for the synchronization of non-identical chaotic systems coupled through a unique scalar signal is based on the design of the slave system as an observer for the master and experimental results supporting our considerations are also provided. The proposed strategy makes use of the negative feedback scheme. The basic idea arose from the evaluation of similar dynamics in two non-identical circuits: the dissipative non-autonomous circuit introduced in Sect. 3.10 and the Duffing oscillator, discussed in Sect. 3.8. Even if the dynamics are different, quantitative conditions can be derived to assess the possibility of achieving synchronization. In order to investigate also the case of non-identical circuits that differ only in a parameter value, the synchronization of two Chua's circuits in different chaotic regions (that is, a single-scroll and a double-scroll attractor) is considered and experimental results given.

5.2.1 Synchronization of Two Chaotic Circuits with Structural Differences

This section deals with the problem of synchronizing two chaotic circuits affected by structural differences. The Duffing oscillator, introduced in Sect. 3.8, and the nonlinear non-autonomous PWL circuit, described in Sect. 3.10, have been considered. The similar fractal dimensions of their attractors (i.e. $d_{DUF} = 1.902$ and $d_{OSC} = 1.903$) calculated through the Grassberger-Procaccia algorithm [18], as well as their similar dynamical behavior, suggest that the two circuits may be synchronized.

The two systems are coupled in a negative feedback scheme. The dissipative non-autonomous circuit is the slave system, so that we only report its equations. They have been modified with the inclusion of terms proportional to the error, which is built by considering the linear combination of both state variables. Hence, the equations of the coupled slave system will read as follows:

$$\begin{cases} \dot{x}_s = y_s - a_1 x_s + k_1 e \\ \dot{y}_s = -b x_s - a_2 y_s + \sin(wt) + s(x_s) + k_2 e \end{cases} \qquad (5.4)$$

where $e = x_m + y_m - x_s - y_s$. Parameter values are chosen in order to ensure that both dynamical systems show a chaotic motion, i.e., $\delta = 0.25$, $\gamma = 0.3$, $b = 1$, $w = 1$. We consider the following values of the observer gains: $k_1 = 0.5$ and $k_2 = 20$. For these values numerical simulations and experimental results show that synchronization may be obtained. The two circuits implementing the dynamics reported in Eqs. (3.42) and (5.4) have been realized in laboratory and waveforms generated by the circuits have been acquired by using a data acquisition board (National Instruments AT-MIO 1620E) with a sampling frequency $f_s = 200\,\text{kHz}$ for $T = 5\,\text{s}$ (this corresponds to 1000000 samples for each time series). Fixing the coupling gains as indicated, the onset of synchronization is clearly visible in Fig. 5.5, where the waveforms of the state variables acquired from the two circuits are reported.

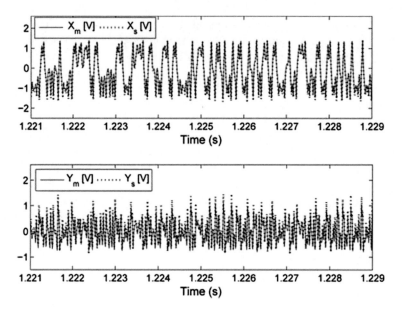

Fig. 5.5 Experimental results showing the onset of synchronization between the non-autonomous PWL oscillator and the Duffing oscillator. Waveforms generated by the two coupled circuits: master system (*continuous lines*), slave system (*dotted lines*)

5.2.2 Synchronization of Two Chaotic Circuits with Parametric Mismatches

In this section, two circuits governed by the same dynamical equations, but with different parameters, are considered. In particular, the two circuits differ for the value of one parameter and then show different behavior (we consider a condition in which both show chaotic behavior, but along different attractors). Let us consider the Chua's circuit as in Eq. (3.5). We consider two different values for α: $\alpha = 9$ (double-scroll chaotic attractor) and $\alpha = 8.6$ (single scroll chaotic attractor).

The two circuits are coupled through a negative feedback scheme in which master and slave equations read as follows:

$$\begin{cases} \dot{x}_m = \alpha_m(y_m - h(x_m)) \\ \dot{y}_m = x_m - y_m + z_m \\ \dot{z}_m = -\beta y_m \\ \dot{x}_s = \alpha_s(y_s - h(x_s)) + k_1 e \\ \dot{y}_s = x_s - y_s + z_s + k_2 e \\ \dot{z}_s = -\beta y_s + k_3 e \end{cases} \tag{5.5}$$

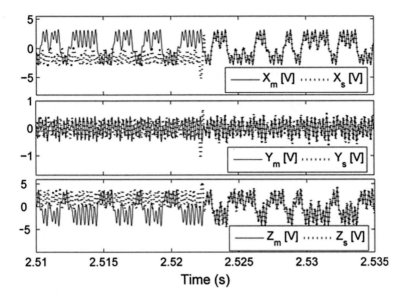

Fig. 5.6 Experimental results showing the onset of synchronization between a Chua's circuit show-ing a single-scroll attractor and a Chua's circuit showing a double-scroll attractor. When the feedback loop is closed at $T = 2.522$ s, the slave system state variables change their behavior. Waveforms generated by the two coupled circuits are: master system (*continuous lines*), slave system (*dotted lines*)

where $e = x_m + y_m - x_s - y_s$ and the following parameters have been considered:

$$\alpha_m = 9, \quad \alpha_s = 8.6, \quad \beta = 14.286, \quad m_1 = \frac{2}{7}, \quad m_0 = -\frac{1}{7} \quad (5.6)$$

The values of coupling gains k_1, k_2, and k_3 have been fixed as $k_1 = -0.26$, $k_2 = 1.33$, and $k_3 = 4.51$. For such values of the observer gains, the two circuits may be synchronized as experimentally demonstrated. In fact, Fig. 5.6 reports the waveforms acquired from the two coupled circuits with the indicated values of the observer gains. In particular, the first part of the experiment is run without connecting the feedback signal, so that the two circuits evolve independently, while in the second part the feedback signal is sent to the second circuit and synchronization is attained. In fact, the slave system evolves along the single-scroll attractor until the feedback loop is closed and then it is forced to follow the double-scroll dynamical behavior of the master system.

5.3 Power Absorption During Synchronization

In this section, a new aspect of the phenomenon of synchronization, emerged dur-ing the experiments on synchronization performed in our laboratory, is dealt with. The new qualitative finding observed is the fact that the power absorbed by the whole system (i.e., the system made by the coupled circuits) is minimum when synchronization is achieved. Different coupling schemes and different dynamical circuits (the Chua's circuit, the Lorenz system, the Rössler circuit as well as some hyperchaotic circuits) have been experimentally investigated, while monitoring the power absorption as the coupling parameter is varied. In all the experiments performed, it has been observed that the minimum power absorbed by the system occurs when the coupling is such to induce a stable synchronous state. In other words, the coupling parameters which guarantee synchronization are those for which the power absorption of the circuits is minimum. Due to the ubiquitous nature of synchronization either in artificial or natural systems, this observation may have important implications in several fields.

We consider here two Chua's circuits coupled through the first state variable in a unidirectional way:

$$\begin{cases} \dot{x}_1 = \alpha[y_1 - h(x_1)] \\ \dot{y}_1 = x_1 - y_1 + z_1 \\ \dot{z}_1 = -\beta y_1 \\ \dot{x}_2 = \alpha[y_2 - h(x_2)] + \sigma(x_1 - x_2) \\ \dot{y}_2 = x_2 - y_2 + z_2 \\ \dot{z}_2 = -\beta y_2. \end{cases} \tag{5.7}$$

Circuit 1 acts as master and circuit 2 as slave. In this case the coupling is implemented using an operational amplifier and several resistors, one of which R_f is assumed to be variable, so that the coupling factor is proportional to R_f.

In our experiments, the average power absorption P_d has been measured with respect to different values of the coupling factor. In particular, the average power absorption has been evaluated separately for each circuit, by measuring the current absorbed by each circuit and the voltage supply provided by the voltage supply generator.

Figure 5.7 shows the power absorption P_d with respect to R_f for both circuits and the average synchronization error E_s, defined as

$$E_s = \langle \sqrt{(x_1 - x_2)^2 + (y_1 - y_2)^2 + (z_1 - z_2)^2} \rangle \tag{5.8}$$

The two circuits do synchronize for $R_f > 1600\ \Omega$. It can be noticed that the power absorption at the master circuit (dashed line) remains constant, while the dissipated power at the slave circuit is high when the two circuits are not synchronized, until it reaches a minimum value when the two circuits do synchronize.

As a second example, we consider two Chua's circuits coupled through the first state variable in a diffusive bidirectional way. As we mentioned previously, this coupling is implemented by inserting a resistor (indicated in the following as R_c)

Fig. 5.7 Synchronization error E_s and power absorption in two unidirectionally coupled Chua's circuits with respect to different values of the coupling resistor R_f

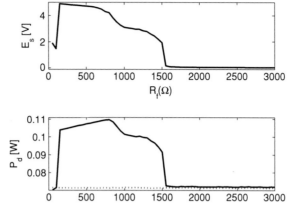

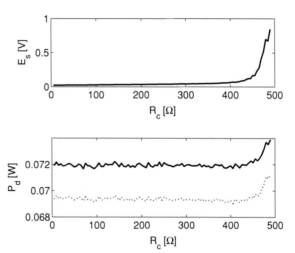

Fig. 5.8 Synchronization error E_s and power absorption in two diffusively coupled Chua's circuits with respect to different values of the coupling resistor R_d

connecting the two capacitors associated with the two corresponding state variables. In terms of dimensionless equations, the system is described by:

$$\begin{cases} \dot{x}_1 = \alpha[y_1 - h(x_1)] + \sigma(x_2 - x_1) \\ \dot{y}_1 = x_1 - y_1 + z_1 \\ \dot{z}_1 = -\beta y_1 \\ \dot{x}_2 = \alpha[y_2 - h(x_2)] + \sigma(x_1 - x_2) \\ \dot{y}_2 = x_2 - y_2 + z_2 \\ \dot{z}_2 = -\beta y_2 \end{cases} \tag{5.9}$$

where in this case the coupling factor k is proportional to the inverse of the coupling resistor, i.e., $\sigma \propto \frac{1}{R_c}$.

Figure 5.8 reports the synchronization error E_s and the power absorption P_d for both circuits as a function of the parameter R_c. The power absorption P_d for both coupled systems decreases monotonically when the coupling resistor is decreased (which corresponds to increasing the coupling k in Eq. (5.9)). In correspondence, the synchronization error decreases when the coupling factor increases. The synchronization error is close to zero when the power absorption is minimum. The synchronization error is not exactly zero, because circuits are nonidentical due to parametric tolerance, and for the same reason there is a slight difference in the power absorption of the two circuits.

The same qualitative results discussed in these two examples have been observed in many other synchronization experiments. Actually, in all of our experiments we have not found a counter example. This shows how power absorption may be a universal signature of synchronization.

References

1. Pikovsky A, Rosenblum M, Kurths J (2001) Synchronization—a universal concept in non-linear science. Cambridge University Press, Cambridge
2. Boccaletti S, Kurths J, Osipov G, Vallardes DL, Zhou CS (2002) The synchronization of chaotic systems. Phys Rep 366:1–101
3. Kapitaniak T (1994) Synchronization of chaos using continuous control. Phys Rev E 50:1642–1644
4. Amritkar RE, Gupte N (1993) Synchronization of chaotic orbits: the effect of a finite time step. Phys Rev E 47:3889–3895
5. Kocarev L, Parlitz U (1995) General approach for chaotic synchronization with applications to communication. Phys Rev Lett 74:5028–5031
6. Parlitz U, Kocarev L, Stojanovski T, Preckel H (1996) Encoding messages using chaotic synchronization. Phys Rev E 53:4351–4361
7. Guemez J, Matias MA (1995) Modified method for synchronizing and cascading chaotic systems. Phys Rev E 52:R2145–R2148
8. Chua LO, Itoh M, Kocarev L, Eckert K (1993) Chaos synchronization in Chua's circuit. In: Madan R (ed) Chua's Circuit: a Paradigm for Chaos. World Scientific, Singapore, pp 309–324
9. Pecora LM, Carroll TL (1998) Master stability functions for synchronized coupled systems. Phys Rev Lett 80:2109–2112
10. Boccaletti S, Latora V, Moreno Y, Chavez M, Hwang D-U (2006) Complex networks. Phys Rep 424:175–308
11. Buscarino A, Fortuna L, Frasca M, Sciuto G (2009) Chua's circuits synchronization with diffusive coupling: new results. Int J Bifurcat Chaos 19(9):3101–3107
12. Stefanski A, Kapitaniak T, Brindley J (1996) Dynamics of coupled Lorenz systems and its geophysical implications. Physica D 98:594–598
13. Osipov G, Pikovsky A, Rosenblum M, Kurths J (1997) Phase synchronization effects in a lattice of nonidentical Rössler oscillators. Phys Rev E 55:2353–2361
14. Yassen M (2005) Chaos synchronization between two different chaotic systems using active control. Chaos, Solitons Fractals 23:131–140
15. Ping Z, Yu-Xia C (2007) Realization of generalized synchronization between different chaotic systems via scalar controller. Chin Phys 16:2903–2907
16. Yau H, Yan J (2008) Chaos synchronization of different chaotic systems subjected to input nonlinearity. Appl Math Comput 197:775–788

17. Sarasola C, Torrealda FJ, D'Anjou A, Moujahid A, Grana M (2003) Feedback synchronization of chaotic systems. Int J Bifurcat Chaos 13:177–191
18. Strogatz SH (1994) Nonlinear dynamics and chaos: with applications to physics, biology, chemistry and engineering. Perseus Books, Reading

Conclusions

This book provides a simple and complete guide for designing and running experiments with chaotic circuits. We have first shown a series of four fundamental chaotic circuits which have been designed by exploiting the peculiar behavior of some electronic components. We have then illustrated a procedure that allows to realize a chaotic circuit which obeys to the same equations of the mathematical model of an arbitrary nonlinear system and, then, a set of examples of circuits built by following this approach. We have depicted two ways for the implementation in the laboratory: the use of discrete components or the use of analog programmable devices such as FPAA. Examples of control and synchronization experiments with two coupled chaotic circuits have been also discussed.

We remark that the proposed circuits are all low cost and easy to realize. All the schemes have been tested so that they can be immediately realized in a laboratory equipped with low-cost instrumentations. Most of the experiments in fact just need a power supply and an oscilloscope, while others also need a waveform generator. Additionally, the procedure introduced also permits to realize circuits from mathematical models not included in the text, so that there is no limit to the possibilities for impressive experiments, applications, and design of chaotic circuits and chaos-based sensors, devices, and control systems.

Realizing nonlinear electronic circuits is today of outstanding importance in the perspective of performing relevant experiments and conceiving new devices, looking at discrete analog circuit design and control as an art, a science, and a fascinating scientific path. Therefore, the manuscript addresses research people toward experiments in electronics, but not only, emphasizing the role of a new professional profile, the analog engineer.

A. Buscarino et al., *A Concise Guide to Chaotic Electronic Circuits*,
SpringerBriefs in Applied Sciences and Technology,
DOI: 10.1007/978-3-319-05900-6, © The Author(s) 2014

So, now, it is your turn to make your own experiments with nonlinear circuits and chaos!

Observations and comments sent to our email addresses will be welcome.

Arturo Buscarino, arturo.buscarino@dieei.unict.it
Luigi Fortuna, luigi.fortuna@dieei.unict.it
Mattia Frasca, mattia.frasca@dieei.unict.it
Gregorio Sciuto, gregorio.sciuto@dieei.unict.it

Index

A. Buscarino et al., *A Concise Guide to Chaotic Electronic Circuits*,
SpringerBriefs in Applied Sciences and Technology,
DOI: 10.1007/978-3-319-05900-6, © The Author(s) 2014